# VACCINES AND WORLD HEALTH

# VACCINES AND WORLD HEALTH

## Science, Policy, and Practice

PAUL F. BASCH

New York   Oxford
OXFORD UNIVERSITY PRESS
1994

Oxford University Press

Oxford   New York   Toronto
Delhi   Bombay   Calcutta   Madras   Karachi
Kuala Lumpur   Singapore   Hong Kong   Tokyo
Nairobi   Dar es Salaam   Cape Town
Melbourne   Auckland   Madrid

and associated companies in
Berlin   Ibadan

Copyright © 1994 by Oxford University Press, Inc.

Published by Oxford University Press, Inc.,
200 Madison Avenue, New York, New York 10016

Library of Congress Cataloging-in-Publication Data
Basch, Paul F., 1933–
Vaccines and world health : science, policy, and practice / Paul F. Basch.
p. cm.   Includes bibliographical references and index.
ISBN 0-19-508532-9
1. Vaccination—Developing countries.
2. Vaccines—Biotechnology.
I. Title.   [DNLM: 1. Vaccines   2. Biotechnology.   QW 805 B298v   1994]
RA638.B37   1994   614.4'7'091724—dc20   DNLM/DLC
for Library of Congress   93-23283

1 3 5 7 9 8 6 4 2

Printed in the United States of America
on acid-free paper

# Preface

Since the inception of biotechnology as a scientific and commercial enterprise in the mid-1970s, many people have been attracted to the prospect that its products might bring about rapid improvements in human health, particularly in the poorest countries of the world. This hope, combined with growing technical capability, has provided the impetus for new thinking about the design, production, transfer, evaluation, and deployment of the physical products of biotechnology and of their underlying methodologies.

Such decisions would be relatively straightforward if based solely on conventional criteria such as safety and efficacy. However, health is so intimately entangled with other aspects of life that anyone who hopes to make a difference in the real world must acknowledge the influence of communications, culture, economics, education, environment, equity, politics, population, and poverty, at the least.

Conceived as a study to evaluate the health potential of existing and anticipated biotechnologic tools, this book has evolved during the process of writing into a discussion that places more emphasis on context and policy, and the original technical chapters have become appendixes to the main text. I decided to concentrate on preventive and diagnostic technologies, particularly vaccines intended for the infectious, vectorborne, and parasitic diseases so widespread in the developing countries. My intention has been to expose the pathway that such products must take to achieve their goal of improving human health, to identify the various groups having a stake in the process, and to portray the concerns and risks facing each.

This book is intended for a variety of readers, coming from different backgrounds and engaged in apparently disparate pursuits:

1. The first group consists of the laboratory "bench" investigators, primarily in industrialized countries, who conceive and develop new products intended to promote health and control disease, who have mastery over the concepts and techniques of biotechnology but little personal experience in the field in developing nations.

2. A second intended audience has the reverse orientation. These are the persons responsible for planning and implementing public health and disease control programs in developing countries. Intimately familiar with the cultural, economic, political, and epidemiologic situation within their own jurisdictions, they may not be so well acquainted with the technical features of modern biomedical technology about which they will need to make decisions.
3. A third group, more varied, is made up of the many people who participate in the long process of taking a product ''from the bench to the bush.'' Included here are investors, managers, engineers, administrators, epidemiologists, and other intermediaries who finance, fabricate, evaluate, and distribute the products of biomedical innovation, as well as those officials responsible for oversight and regulation.
4. Finally there are the students of medicine, public health, public policy, economics, international relations, anthropology, and other disciplines that have an interest in technology transfer and in international health.

My self-appointed task has been a daunting one, requiring greater experience, knowledge, wit, and skill than I can fairly claim. The experience of others, recorded in an abundant literature, has been the source of much information. Here and there in the text I have inserted brief excerpts from these writings, to represent in their own words the ideas of experts and commentators. Still, this book is likely to contain errors of fact or interpretation. I wish I knew what and where they are.

I sincerely appreciate the confidence and support of the National Library of Medicine in the form of a writing grant that long ago receded into history but established the nucleus of this volume and a stimulus for continued work. Lorela Fajardo, Lawrence McEvoy, Tung Nguyen, and Brian Scullion have made special bibliographic efforts on behalf of this project. I am grateful to the scientific and professional colleagues who have read and commented on various drafts of the manuscript, discussed issues with me, or provided other information. John Barton, Sean Bourke, Alexandra Fairfield, Stanley Foster, Stephen Jarrett, Robert Kim-Farley, P-H Lambert, Lindsay Martinez, Jaime Martuscelli, Gerald Rosenthal, Edwin Rock, Franz Rosa, and Paul Zukin have been particularly helpful, but as their advice has not always been incorporated, the responsibility for errors rests with me alone. I thank associates in Brazil, Egypt, India, Indonesia, Malaysia, Mexico, Pakistan, Zimbabwe and other countries in which I have had the opportunity to observe the technology transfer process firsthand. And special thanks go to my wife, Natalicia, and sons, Richard and Daniel, for their prolonged forbearance through innumerable hours of distraction.

*Stanford, California*                                                                                     P. F. B.
*March 1993*

# Contents

# Abbreviations and Acronyms

| | |
|---|---|
| ACIH | Agency for Cooperation in International Health (Japan) |
| ACIP | Immunization Practices Advisory Committee (USA) |
| ACMR | Advisory Committee on Medical Research (WHO) |
| ADB | Asian Development Bank |
| AfDB | African Development Bank |
| AHRTAG | Appropriate Health Resources and Technologies Action Group (UK) |
| A.I.D. | United States Agency for International Development (USAID) |
| AIDAB | Australian International Development Assistance Bureau |
| AIDS | Acquired immunodeficiency syndrome |
| AR | Attack rate |
| ARI | Acute respiratory infection |
| ALRI | Acute lower respiratory infection |
| BCG | Bacille Calmette-Guérin (TB vaccine) |
| BI | Biotechnologic innovation |
| BINAC | Biosciences Information Network for Latin America and the Caribbean |
| CBA | Cost-benefit analysis |
| CCCD | Combatting childhood communicable diseases (USAID) |
| CDC | Centers for Disease Control and Prevention, US Public Health Service |
| CDD | Control of Diarrheal Disease (WHO) |
| CDR | Division of Diarrheal and Acute Respiratory Disease Control (WHO) |
| CD&RI | Contraceptive development and research in immunology |
| CEA | Cost-effectiveness analysis |

| | |
|---|---|
| CEU | Clinical Epidemiology Unit (INCLEN) |
| CGIAR | Consultative Group on International Agricultural Research |
| CGVD | Consultative Group on Vaccine Development (USPHS) |
| CHW | Community health worker |
| CIDA | Canadian International Development Agency |
| CIOMS | Council for International Organizations of Medical Sciences |
| CMR | Child mortality rate; Crude mortality rate |
| COBIOTECH | Scientific Committee on Biotechnology (UNESCO) |
| COGENE | Committee on Genetic Experimentation (UNESCO) |
| CPHC | Comprehensive primary health care |
| CVI | Children's Vaccine Initiative |
| dAb | Single domain antibody |
| DANIDA | Danish International Development Agency |
| DCD | Department of Cooperation for Development (Italy) |
| DGDC | Directorate General for Development Cooperation (Netherlands) |
| DOC | Department of Commerce (United States) |
| DPT | Diphtheria, pertussis, tetanus (same as DTP) |
| DTP | Diphtheria, tetanus, pertussis (same as DPT) |
| EC | European Community |
| E-E | Edmonston-Enders vaccine (measles) |
| EIA | Enzyme immunoassay |
| EIS | Epidemic Intelligence Service (CDC) |
| ELISA | Enzyme linked immunosorbent assay |
| EMBO | European Molecular Biology Organization |
| EPA | Environmental Protection Agency (United States) |
| EPI | Expanded Programme on Immunization (WHO) |
| E-Z | Edmonston-Zagreb vaccine (measles) |
| FDA | Food and Drug Administration (United States) |
| FETP | Field Epidemiology Training Program (see EIS) |
| FHW | Family health worker |
| FIC | Fully immunized child; Fogarty International Center (NIH) |
| FIFRA | Federal Insecticide, Fungicide, and Rodenticide Act (United States) |
| FINNIDA | Finnish International Development Agency |
| GAG | Global Advisory Group (EPI) |
| GATT | General Agreement on Tarrifs and Trade |
| GMP | Good manufacturing practices |
| GNP | Gross national product |

| | |
|---|---|
| GOBI | Growth monitoring, oral rehydration, breastfeeding, immunization |
| GOBI/FFF | GOBI plus family planning, food production, female education |
| GPA | Global Programme on AIDS (WHO) |
| GTZ | Gesellschaft für Technische Zusammenarbeit (Agency for Technical Cooperation) (Germany) |
| HBV | Hepatitis B virus |
| HFA | Health for all (WHO) |
| HFA2000 | Health for all by the year 2000 (WHO) |
| HIV | Human immunodeficiency virus |
| IAEA | International Atomic Energy Agency |
| IBI | Initiative for Biotechnology Implementation (WHO/TDR) |
| IBRD | International Bank for Reconstruction and Development (World Bank) |
| ICGEB | International Centre for Genetic Engineering and Biotechnology (UNIDO) |
| ICRO | International Cell Research Organization (UNESCO) |
| ICSU | International Congress of Scientific Unions (UNESCO) |
| IDRC | International Development Research Centre (Canada) |
| ICDDR,B | International Centre for Diarrhoeal Disease Research, Bangladesh |
| IDB | Inter-American Development Bank |
| IIF | Indirect immunofluorescence |
| IFPMA | International Federation of Pharmaceutical Manufacturers Associations |
| IMF | International Monetary Fund |
| IMR | Infant mortality rate |
| INCLEN | International Clinical Epidemiology Network (Rockefeller Foundation) |
| IOBB | International Organization for Biotechnology and Bioengineering |
| IPM | Integrated pest management |
| IPR | Intellectual property rights |
| IPV | Inactivated poliomyelitis vaccine (same as KPV) |
| IRB | Institutional review board |
| ISCOM | Immunostimulating complex |
| IRTC | Immunology Research and Training Centre (WHO, Lausanne) |
| JAIDO | Japan International Development Organization |
| JICA | Japan International Cooperation Agency |
| KPV | Killed poliomyelitis vaccine (same as IPV) |
| LDC | Less developed country |
| MAb | Monoclonal antibody |
| MCH | Maternal and child health |
| MIRCEN | Microbiological Resource Centers (UNESCO) |

| MMR | Maternal mortality rate; Measles, mumps, and rubella vaccine |
| MMWR | Morbidity and Mortality Weekly Report (CDC, USPHS) |
| MOH | Ministry of Health; Medical Officer of Health |
| NEPA | National Environmental Policy Act (United States) |
| NGO | Non-governmental organization |
| NIH | National Institutes of Health (USA) |
| NNT | Neonatal tetanus |
| NORAD | Norwegian Agency for International Development |
| NVAC | National Vaccine Advisory Committee (USA) |
| NVP; NVPO | National Vaccine Program; National Vaccine Program Office (USA) |
| NNMR | Neonatal mortality rate |
| NNT | Neonatal tetanus |
| OAS | Organization of American States |
| ODA | Overseas Development Administration (United Kingdom); Official development assistance |
| OECD | Organization for Economic Cooperation and Development |
| OIH | Office of International Health (USA) |
| OPRR | Office of Protection from Research Risks (USPHS) |
| OPV | Oral poliomyelitis vaccine |
| OSTP | Office of Science and Technology Policy (United States) |
| OR | Odds ratio |
| ORS | Oral rehydration solution; oral rehydration salts |
| ORT | Oral rehydration therapy |
| OTA | Office of Technology Assessment (US Congress) |
| PAHO | Pan American Health Organization |
| PASB | Pan American Sanitary Bureau |
| PATH | Program for Appropriate Technology in Health |
| PDG | Product Development Group (CVI) |
| PEM | Protein-energy malnutrition |
| PHC | Primary health care |
| PNN | Postneonatal |
| PVD | Programme for Vaccine Development (WHO/UNDP) |
| PVO | Private voluntary organization |
| QA | Quality assurance |
| QALY | Quality adjusted life year |
| QC | Quality control |
| RAC | Recombinant DNA Advisory Committee |

| | |
|---|---|
| RCT | Randomized clinical trial |
| R&D | Research and development |
| RR | Relative risk |
| RSV | Respiratory syncytial virus |
| SAGE | Scientific Advisory Group of Experts (PVD) |
| SAP | Structural adjustment policy |
| SCP | Single cell protein |
| SDC | Swiss Development Corporation (Switzerland) |
| SIDA | Swedish International Development Authority |
| SIREVA | Sistema Regional de Vacunas [Regional Vaccine System] in the Region of the Americas |
| SPHC | Selective primary health care |
| S&T | Science and technology |
| STD | Sexually transmitted disease |
| TA | Technology assessment; Technical assistance |
| TAG | Technical advisory group |
| TB; TBC | Tuberculosis |
| TBA | Traditional birth attendant |
| TDR | Special Programme for Research and Training in Tropical Diseases (UNDP/ World Bank/WHO) |
| TDV | Transdisease vaccinology (PVD) |
| TECHCOM | Technical consultative meeting (EPI) |
| TECHNET | Technical Network for Logistics in Health (WHO/UNICEF) |
| TF | Technology forecasting |
| TNC | Transnational corporation |
| TOSCA | Toxic Substances Control Act (United States) |
| TT | Tetanus toxoid |
| UCI | Universal childhood immunization |
| UN | United Nations |
| UNCTAD | United Nations Conference on Trade and Development |
| UNCSTD | United Nations Conference on Science and Technology for Development |
| UNEP | United Nations Environmental Program |
| UNESCO | United Nations Educational, Scientific, and Cultural Organization |
| UNICEF | United Nations Children's Fund |
| UNIDO | United Nations Industrial Development Organization |
| UNDP | United Nations Development Programme |
| UNU | United Nations University (Tokyo) |

| | |
|---|---|
| URI | Upper respiratory infection |
| USAID | United States Agency for International Development (A.I.D.) |
| USPHS | United States Public Health Service |
| UV | Ultraviolet |
| VHW | Village health worker |
| VII | Vaccine independence initiative (EPI) |
| WHA | World Health Assembly (WHO) |
| WHO | World Health Organization |
| YHLL | Years of healthy life lost |
| YPLL | Years of productive (or potential) life lost |

# VACCINES AND WORLD HEALTH

# 1

# Concept, Reality, and Risk in Technological Innovation

Throughout history, virtually all human activities have had their ups and downs: arts and religions have waxed and waned; economies have grown and contracted; nations and empires have prospered and decayed. Since the earliest settlements and civilizations, perhaps the only human activities that have advanced in an uninterrupted, if sometimes unsteady, forward direction are science and technology.

The application of technology helps human beings to:

1. Locate, identify, gather, transport, and modify useful natural materials
2. Protect themselves from hostile aspects of the physical and biotic environment
3. Express their creativity
4. Escape from the limitations imposed by their physical strength and their five senses: to see farther or smaller than with the unaided eye; to hear music not being made by living players; to communicate and transmit information across continents and decades; to travel faster; to eat foods not freshly harvested or killed; to prevent and cure certain ailments, and so on

## Technology and Development

Technology has spurred economic development and productivity from the earliest days until the present, and will continue to do so for the foreseeable future. However, technology does not grow uninterruptedly in any one site. Great ancient civilizations in China, Egypt, India, Middle and South America, Mesopotamia, and Africa developed in regions now considered part of the "Third World." Despite their distinguished heritage, such countries are characterized by a deficiency of science and technology by current world standards.

## Technology and Health

Health-related technology in ancient times lacked a scientific basis and was applied empirically in cases of injury or illness of humans or domesticated animals. Much has been of enduring value. Herbal cures cataloged by ancient Chinese and Ayurvedic medical writers are still in use, as are derivatives of tree barks—cinchona for fevers and ipecac for diarrheas—used by early inhabitants of South America. In parts of the Middle East where cutaneous leishmaniasis is prevalent, it was observed in ancient times that usually a single "oriental sore" occurred during one's lifetime. Consequently, material was transferred from an active lesion of one person to an inconspicuous area of another: a prophylactic inoculation on the buttocks of a young child would prevent a disfiguring facial lesion later in life. In a more hazardous example, the practice of variolation arose in ancient Turkey, China, and perhaps independently elsewhere. To induce immunity, a powder made from a pustule from a patient with a mild case of smallpox was sniffed through the nose. Often successful, the procedure sometimes resulted in a fatal case of smallpox. Variolation was taught to American colonists by their African slaves, and became fairly widespread in the Western Hemisphere before being supplanted by vaccination in the late 1790s.

The origins of biotechnology can be traced to the earliest civilizations. Domestication and selective breeding of animals and plants was carried out with deliberate intent all around the ancient world, inaugurating both agriculture and husbandry. The control of fermentation to make beers and wines, yogurt, cheeses, soya sauce, leavened bread, and other products was discovered independently at many different times and places.

These technologic developments, made with little or no comprehension of the underlying genetic, pharmacologic, or physiologic principles, have achieved widespread adoption. Their mobility and popularity suggest that a useful technology can be transplanted to a country that lacks scientific capability, so long as the necessary materials and enabling conditions are available. However, as I have pointed out elsewhere (Basch 1993a):

> Those who do things that they do not understand have little control over their procedures and remain at the mercy of forces that they can not entirely dominate. True mastery over a technology demands comprehension of underlying scientific principles. Transplanted technology without science is inviable in the long term and can not lead to an indigenous capacity to solve problems independently.

*Biomedical knowledge in ancient times.* Science, in the sense of astute observations about recurring relationships that lead to the explanation of natural phenomena, has always been closely associated with health. Lacking thermometer and microscope, Hippocrates (ca. 460–380 B.C.) accurately distinguished tertian and quartan fevers. Today these forms of malaria are known to be caused by different species of *Plasmodium* parasites, about which Hippocrates knew nothing. There is evidence that during the same era, some African peoples knew that tse-tse flies transmit sleeping sickness, and that mosquitoes carry malaria.

*Biomedical knowledge in modern times.* Only in fairly recent times, perhaps since Lind's demonstration of the nutritional cause of scurvy and Jenner's introduction of vaccination in the late eighteenth century, has western biomedical science developed as an independent discipline. From that era onward there has been sustained optimism about the imminence of medical advances. Philadelphia's Dr. Benjamin Rush, of Revolutionary War fame, wrote in 1815:

> Could we lift the curtain of time which separates the year 1847 from our view, we should see cancers, pulmonary consumptions, apoplexies, palsies, epilepsy and hydrophobia struck out of the list of mortal diseases and many others which still retain an occasional power over life, rendered perfectly harmless, provided the same number of discoveries and improvements shall be made in the intermediate years, that have been made since the year 1776.

The explosion of knowledge dating from the mid-nineteenth century marks the origin of modern biomedicine. Darwin's intellectual framework, Pasteur's demonstration of the role of microorganisms in fermentation, and Buchner's discovery that enzymes isolated from yeast could catalyze conversion of sugar to alcohol in the absence of living cells opened the floodgates of discovery.

Knowledge of pathology, bacteriology, physiology, nutrition and many other "classical" biomedical disciplines has combined with advances in chemistry, optics, and other fields to form the basis of a mature biomedical science. Mastery of fermentation technology led to industrial-scale production of acetone, glycerol, and other products, and, starting in the late 1940s, to the elaboration of the antibiotic industry. The "new" biotechnology, made possible in 1953 with the Watson-Crick model of DNA structure, has tied together tissue and cell culture, cell fusion technology, microbial fermentation, immunogenetics, and the manipulation of heredity.

By 1992 more than 1,100 biotechnology companies were operating in the United States alone. Sales of biotechnology products from the United States, United Kingdom, Japan, Germany, and France were more than US$8 billion, with an expectation of $100 billion in worldwide sales by the year 2000 (Burrill and Roberts 1992).

The rapid advances of recent decades have stimulated universal interest and excitement. The prospect of applying novel and specialized techniques and products to the health needs of the poorer countries has attracted many adherents in developing and industrial nations alike. Among potential applications of biotechnology, few have been as appealing as the development of novel vaccines for prevention of infectious diseases.

## Necessity and Invention

The old aphorism, "Necessity is the mother of invention," is well entrenched in everyday life, as humans have always devised ways to do things that needed doing. Curiosity and creativity are haphazard, unpredictable, and unplannable. Occasionally, flashes of brilliance and ingenuity derive from seemingly unrelated antecedents, as when Alexander Fleming saw a petri dish in which bacterial growth was inhibited, and conceptualized the idea of antibiotics. More commonly, basic research, which is really disciplined curiosity, leads gradually and haltingly to new knowledge and insights.

Sometimes criticized as unnecessary and wasteful, basic research is a wellspring for unimagined consequences. Modern science is the product not of directed "Research and Development," but of the application of organized curiosity, in the guise of basic research, often seemingly unrelated to the nature of the later practical innovation. Many key discoveries in the present century, rewarded by Nobel Prizes, have generated valuable practical applications and contributed to the growth of biotechnology (Table 1-1). It seems no exaggeration to turn the old aphorism on its head and say, with Nobel laureate Arthur Kornberg, that "invention is the mother of necessity" as we become more and more dependent on technological developments that have arisen from innovative thinking.

## Taking Discoveries from the Laboratory to the Field

To all people, the potential for a longer and more productive life, with decreased mortality and relative freedom from preventable illness, presents a strong attrac-

TABLE 1-1.  Some Nobel Laureates and Their Biomedical Discoveries That Have Led to Significant Practical Innovations

| Nobel Laureates | Year | Discovery |
|---|---|---|
| Warburg | 1931 | Respiratory enzyme |
| Fleming, Chain, and Florey | 1945 | Antibiotic effect of penicillin |
| Waksman | 1952 | Streptomycin |
| Enders, Wellers, and Robbins | 1954 | Growth of poliovirus in tissue culture |
| Beadle, Tatum, and Lederberg | 1958 | Genetic regulation; bacteria genetics |
| Ochoa and Kornberg | 1959 | Biosynthesis of RNA and DNA |
| Burnet and Medawar | 1960 | Immunological tolerance |
| Crick, Watson, and Wilkins | 1962 | Molecular structure of nucleic acids |
| Holley, Khorana, and Nirenberg | 1968 | Intepretation of genetic code |
| Delbrück, Hershey, and Luria | 1969 | Genetic structure of viruses |
| Edelman and Porter | 1972 | Chemical nature of antibodies |
| Baltimore, Dulbecco, and Temin | 1975 | Tumor virus and host cell interaction |
| Arber, Nathans, and Smith | 1978 | Restriction enzymes, molecular genetics |
| Benacerraf, Dausset, and Snell | 1980 | Cell surface immunologic determinants |
| Bergström, Samuelsson, and Vane | 1982 | Prostaglandins |
| Jerne, Köhler, and Milstein | 1984 | Monoclonal antibodies |
| Cohen and Levi-Montalcini | 1986 | Growth factors |
| Tonegawa | 1987 | Genetics of antibody diversity |
| Black, Elion, and Hitchings | 1988 | Principles of drug action |
| Bishop and Varmus | 1989 | Origin of retroviral oncogenes |
| Murray and Thomas | 1990 | Organ and cell transplantation |

tion. For the citizens of the poorer countries, no benefit of modernization may be so valued as improved health, particularly for infants and children. Although it is acknowledged that such benefits accompany an improving economy in the sense that "a rising tide lifts all ships," the residents of developing countries may find it difficult to wait for future prosperity to achieve these goals. In fact, the current debt crisis demonstrates the reverse, as dreams of good fortune recede in many parts of the world. Accordingly, individuals, institutions, and governments turn to technology to alleviate, or at least to palliate, a discouraging health situation.

*The relevance of new technology.* Despite their general popularity, new technologies are not universally welcomed. Speaking of Africa, Barry (1989) pointed out that

> the resistance of the rural population to technological innovation is deeply rooted in its attachment of traditional values and structures. This attachment is inherent in the very nature of African peasants; as they generally have no specific training, they are prudent, skeptical and pragmatic. They believe what they can see . . . they have hardly any inclination for novelty or risk. . . . Technologies that are imposed from the outside without consulting the peasants, without appealing to them for co-operation, are as expensive as they are off-putting to disadvantaged population groups.

Even within the tropical medicine profession, some experienced observers view emerging technologies and their advocates with a suspicion that borders on hostility:

> During the past two decades when biotechnology has made so many stunning advances, the health of tropical peoples has worsened. Eradication and control schemes have collapsed. Old, proven therapies have become impotent in the face of drug-resistant organisms. New, affordable, non-toxic chemotherapeutics have not been developed. . . . Expertise has been lost; the last generation of truly experienced "field hands" are leaving the scene, lost to age and disuse. They are being replaced both in the West and in the research centers of the tropics by the "molecular types," more concerned with the exquisite intellectual challenges of modish science than with seeking practical solutions. The razzledazzle and promise of biotechnology have led Third World health officials to expect the quick fix—the malaria vaccine "just around the corner," the genetically altered mosquito that yesterday's press release proclaims will be the last word in controlling vector-borne diseases; and confusing diagnosis with cure, the DNA probe techniques to detect parasites even at clinically insignificant levels. . . . There is an imbalance, a discontinuity between research and reality. This is an imbalance that has inhibited improvement in the health of tropical peoples; but in addition, I believe it has actually contributed to the deterioration of health (Desowitz 1991).

Where is the truth? Clearly, we need an objective evaluation of the probable role of new biomedical technology in relation to the health of the world's people.

*Technology transfer.* Despite the volumes that have been written on the subject, no universally accepted definition exists for "technology transfer." Many authors wisely avoid the term. The technology referred to may be a physical product such as a piece of hardware or biological material, or a specific methodology or technique. Successful transfer requires not only movement of the innovation, but also of its underlying conceptual and functional basis. It may be said that technology transfer (rather than mere exportation and importation of a commodity) has occurred when a discernible permanent difference is induced in a country as a result of having received the technology.

*Incongruous use of transferred technology.* Modern bicycles are built in two configurations. The men's model has an upper horizontal tube extending from under the handlebars to beneath the seat. In the women's bicycle the equivalent tube slopes downward from beneath the seat to the region of the pedal crank. The two patterns were developed, of course, because European men wore pants and women wore dresses. In Malaysia, where bicycles were introduced by the British, it is common to see Chinese women or Malay men riding bicycles. The Chinese women, however, who usually wear the pajama-like samfoo, ride on women's bicycles, and rural Malay men, with their long skirt-like sarongs, always use the men's model, exactly reversing the rationale for the two designs. Technology is not always employed as envisioned by its developers.

*Technology for public health.* In a limited sense a new vaccine or immunodiagnostic for tropical diseases, intended for community-wide use in poor nations, is similar to any other product in that it must be conceived, designed, tested, manufactured, paid for, distributed, and used. However, conventional commercial channels and established economic relationships generally do not apply to the transfer of biotechnologic innovations intended to improve the health of the general public in developing countries. In the conventional commercial transaction, technology transfer is carried out through direct investment, joint ventures, licensing arrangements, or turnkey or service contracts, sometimes combined with technical assistance programs. In contrast, public health innovations typically include a mixture of scientific, social, and humanitarian motives on the part of their developers, with varying expectation of financial profit. Also, the consumer of the product and the purchaser who pays for it are usually very different: one may be an infant in a remote rural area, while the other is a Ministry of Health or a foreign donor agency. A U.S. Congressman has put his finger on a fundamental issue:

> It is possible to conceive of dozens of uses for biotechnology that have little commercial value but obvious social value. The example that comes to mind most quickly is the malaria vaccine. Over 150 million people get malaria each year, but they are among the poorest people in the world. What commercial incentive exists to invest development money if the potential consumers have no buying power (Dingell 1985)?

*Health-related needs in developing countries.* Other than nutrition, psychological and physiological stability, and protection from environmental hazards, it is difficult to specify the health "needs" of an individual, community, nation, or of the developing world in general. Needs are usually defined by the prevailing official health services in terms of

1. Policy (for example, emphasis on child survival)
2. Targets (reduce the incidence of pertussis to 2 cases/100,000 population)
3. Services (provide polio immunization or AIDS testing)

"The desires of the public themselves, and their demands against the system, may or may not be the same as their presumed needs as determined *by* the system, and to the extent that this discrepancy exists it may affect the acceptability of certain health related technologies" (Basch 1993a).

> As a rule, development programmes for target populations are designed and even carried out from the outside by experts who are very often foreign to the environment; the population takes no part in the design, the planning or the implementation of the projects, which are after all supposed to improve its living conditions. . . . Experts arrive in rural zones and decide to carry out such or such a project without even consulting the people who are to benefit by it, without taking into account their genuine needs or their particular economic situation (Barry 1989).

In satisfying consumer wants the health sector is unlike any other product or service. One refrigerator per household is usually enough, and one or perhaps two TV sets. In the education sector, some people will get MDs and PhDs, or maybe both, but even they will eventually stop. By contrast, the demand for personal medical services is not readily saturable. The very ingenuity that we want to stimulate provides a limitless potential for useful procedures and products, each of which generates needs for training, facilities, and financing.

Despite all of the effort expended, it has never been possible to measure, or even to define, health. Accordingly, the assessment of success at meeting presumed health needs is often tenuous and indirect (Table 1–2). This vagueness causes great problems for the health sector, which must compete for the attention, if not the affection, of government officials.

Table 1–3 suggests that new vaccine technology generally takes five to fifteen years to diffuse on a significant scale throughout an industrial society, and probably longer in developing countries.

*The Mount Everest effect.* While the benefits of an innovation may be self-evident to its creators, transfer to developing countries must be preceded by a searching analysis of its intended role. The existence of a proposed remedy does not establish that it fulfills any corresponding "need" in the community. Availability of a vaccine or diagnostic reagent that is safe, efficacious, potent, and stable is clearly a necessary, but not a sufficient, condition for its adoption, nor even for initiating costly field trials.

"Because it is there" may be ample justification to climb a mountain, but does

TABLE 1–2. Measuring Progress in Various Sectors

| Sector | Type | Criteria |
|---|---|---|
| Agriculture | D | Volume, quality, and appropriateness of food and fiber |
| Industry | D | Variety, volume, and quality of products; sales |
| Commerce | D | Bank transactions, employment statistics, tax receipts |
| Education | D | Test achievement; comparison of output to national needs |
| Health | P | Resources: medical schools, hospital beds, health centers, nurses, doctors, etc. |
| | P | Volume of services: immunizations, deliveries in hospitals, admissions, diagnostic procedures, etc. |
| | P | Demographic indicators: infant and other mortality rates, expectation of life, etc. |
| | P | Morbidity indicators: incidence and prevalence of acute and chronic diseases, lost school and work days, etc. |
| | P | Personal health services: consumer satisfaction |
| | P | Public health services: reduction in disease transmission, absence of outbreaks or epidemics, etc. |

D = Direct measure
P = Proxy measure

TABLE 1–3. Time in Years from a Key Discovery in the Development of Selected Vaccines until Government Approval

| Vaccine | Discovery | Years from Discovery | | |
|---|---|---|---|---|
| | | First Human Experiments | Large-Scale Field Trials | Vaccine Approval |
| Polio (Salk) | Propagation of virus in culture (1949) | 3 | 5 (USA) 6 (France) 5 (USSR) | 6 (USA) 8 (France) 7 (USSR) |
| Rubella | Propagation of virus in culture (1962) | 2 | 6 (USA) 7 (France) | 7 (USA) 10 (France) 10 (UK) |
| Hepatitis B | Identification of surface antigen | 1 | 10 (USA) 9 (France) | 11 (USA) 11 (France) 12 (UK) |

*Source:* Petricciani et al. 1989.

not serve as well for the adoption of technologies. Just because something may be technically achievable does not necessarily mean that it is needed to attain a particular goal (see *alternative risk,* Table 1–5). For example, a well-known laboratory researcher on the biology of amebae published the following sentence about amebiasis: "As with other infectious diseases, vaccination would be the most cost-effective approach for prevention." Although this may be so, no evidence was presented, and it seems at least equally plausible that washing one's hands after defecation and before eating is likely to be more effective, less costly, and induce fewer adverse effects than the predicted vaccine.

In another example, Werth (1991) described the aggressive marketing of recombinant human growth hormone in the United States. He showed how the availability of this product was used to redefine a situation formerly considered normal (relative shortness of stature) as a condition requiring medical intervention, thereby "legitimizing the treatment of healthy kids."

Despite all the caveats, studies have shown that biomedical innovations diffuse all over the world more quickly than developments in the nonmedical sphere, such as television or computers. Piachaud (1979) sent questionnaires to health officials in 85 developing countries asking about the adoption of eight modern technologies (examples: ultrasonic fetal examination, cardiac catheterization, cobalt bomb therapy). Of the 40 countries that responded, almost all had adopted some, and three had incorporated all of the eight fairly costly innovations.

*Cosmetic versus effective biomedical innovations.* The most difficult task is differentiating between those technologies that will fulfill a real need and be cost-effective in the local context, and cosmetic technologies that may be appealing but do not really improve the public health. The following checklist provides a preliminary screen to assess whether an innovation may be usefully transferred to a developing country. Novel biomedical technologies intended for primary health care applications in developing countries should embody these characteristics:

1. *Efficacy and safety.* The desired result is obtained routinely, without harm to the individual, community or environment.
2. *Significance.* The condition for which the technology is intended is important enough to make the effort worthwhile for all concerned.
3. *Performance.* The technology functions properly under local conditions of temperature, dust, electrical power, water, spare parts, and (un)trained personnel.
4. *Acceptability.* The technology is physiologically tolerable and culturally satisfactory to the people for whom it is intended.
5. *Affordability.* The cost is within the available resources of the community or of the local or national health services.
6. *Transferability.* The technology is operable by local people who can understand it, learn to use it, maintain it, and if feasible, to make it.
7. *Sustainability.* The technology continues to be used after the foreign experts and donors have left and the field trial has ended.

An interesting nonbiotechnologic example of effective application of a health technology is the use of ultrasonography to determine the extent of morbidity from the parasitic disease schistosomiasis. In this disease, severe changes can occur in the liver, portal veins, urinary bladder, and other organs. These conditions are difficult to evaluate without extensive and costly examination in a hospital. Ultrasound provides a rapid, noninvasive means of physical diagnosis without a laboratory, and can distinguish between schistosomal fibrosis and posthepatic cirrhosis. Portable ultrasound machines, about the size of small computer monitors, powered by mains or portable generators, have been carried into villages in Sudan

and Egypt (Homeida et al. 1988; Abdel-Wahab et al. 1990) and other endemic
areas such as Cameroon, Niger, and Tanzania.

## The Case of Immunization

The history of immunization has been summarized by Dunlop (1988) and is discussed also in Appendix B in relation to vaccine technology. Encouraged by the
outstanding success of the worldwide Smallpox Eradication Campaign, many people have concluded that immunization offers the clearest path to individual protection from, and control of tropical diseases. Schild and Assaad (1983), for example, believed that

> mankind is now on the threshold of a new era in the technology of vaccine devel-
> opment and production, which stems from important advances in biotechnology, in
> particular recombinant DNA and cell fusion techniques. It offers hope of producing
> vaccines for many of the diseases that are yet uncontrolled and also of developing
> vaccines that are more effective, safer, and more cost-effective than those in cur-
> rent use.

Referring specifically to parasitic diseases of humans, for which immunization
remains unavailable, Soulsby (1982) asked rhetorically why vaccines are needed
in the presence of chemotherapeutic drugs, then answered his own question:

> Parasitic disease tends to be the most common in the areas of the world where the
> population can least afford the available antiparasitic drugs. Further, effective che-
> motherapy may require repeated treatments, adding to the cost but also posing
> logistical problems of availability of drugs, the education of people in their use,
> the administration of them if parenteral treatment is required and, in some cases,
> resistance to the chemotherapy may be developed by the parasite. This is particu-
> larly so in the case of the protozoa. There is no good evidence that chemotherapy
> alone can cope with the widespread parasitic diseases of the developing world and
> vaccines, with their need for limited numbers of treatments, offer great advantages.

The Institute of Medicine of the U.S. National Academy of Sciences undertook
a lengthy study of the potential for vaccine development. Their report, issued in
1986 as two large volumes dealing with Diseases of Importance in the United
States, and in Developing Countries, was a landmark in this subject (Institute of
Medicine 1986). The panel concluded that the greatest potential benefits to world
health would be made by implementation of candidate vaccines against the following pathogens: *Streptococcus pneumoniae,* which causes pneumonia, meningitis,
and otitis media through the world; *Plasmodium* species, the agents of malaria,
particularly *P. falciparum,* which causes malignant tertian malaria; rotaviruses,
responsible for a high burden of childhood diarrheas; *Salmonella typhi,* the pathogen responsible for typhoid fever (specifically the Ty21a oral vaccine); and *Shigella* species, bacterial agents of bacillary dysentery and diarrheas.

Of lesser potential significance were vaccines against hepatitis B virus; *Haemophilus influenzae* type b; the diarrhea-causing bacteria *Escherichia coli;* group

A *Streptococcus; Mycobacterium leprae,* the agent of leprosy; *Vibrio cholerae,* which causes cholera; respiratory syncytial virus; parainfluenza virus; or a new rabies virus. Other agents were put into a still lower category.

## Basic Parameters of Immunization

Many pathogens, including the agents of measles, rubella, pertussis, tetanus, and tuberculosis, are so common that all countries can be considered endemic areas. Others, such as the organisms that cause malaria or schistosomiasis, are widespread, but not ubiquitous, in tropical areas, and some, such as Chagas' disease or African sleeping sickness, are geographically circumscribed. Within countries and populations, certain diseases are more prevalent among limited population segments. Therefore, each kind of immunization will have its particular strategy and epidemiologic goal.

## Clients of Immunization

Residents of endemic areas are the most numerous potential consumers of vaccine, but not the most lucrative market. The target population consists of individuals susceptible to infection and disease, primarily children (with the possible exception of potential AIDS and contraceptive vaccines). Immunizations may be given on an ongoing basis in clinics and other health facilities, in mass campaigns, in schools, or by other strategies, as discussed in chapter 5.

Travelers or migrants to endemic areas are also candidates for immunization. A. Internal travelers, groups of people ethnically similar to endemic area residents, may enter for economic, security, or other reasons and remain for varying periods. They should be protected from infection and disease. B. Nonimmune temporary visitors to endemic areas, while not a large fraction of those exposed, are likely to be a significant market for the producers and financial backers of a vaccine. Their exposure may be transient or prolonged. Prominent among such groups are the military, social service volunteers such as Peace Corps, missionaries, businessmen, workers in extractive industries, and tourists. Protection of visitors may be important to a poor country that depends on income from tourism.

## Functions of Immunization

The main reasons for immunization against infectious agents are:

1. To change the host-parasite relationship in the immunized individual. A. *Primary prevention* blocks infection by killing incoming pathogens or rendering them incapable of attachment, organ or cell penetration, development, and proliferation. B. *Secondary prevention* reduces pathology and disease, whether acute or chronic, after infection has occurred.

2. To protect others directly. The best example of this altruistic function is immunization of children against rubella virus. The aim is to protect the future fetus by preventing maternal rubella infection during pregnancy. Although the direct target of immunization is prepubescent girls, children of both sexes may be immunized to mini-

mize transmission of rubella virus in the community. To the extent that maternal anti-bodies, transmitted through the placenta or via colostrum and milk, protect the young infant after birth, immunization of pregnant women with tetanus toxoid or other anti-gens serves a similar purpose.

3. To protect others indirectly by reducing transmission. Reducing transmission is equivalent to controlling a disease. Control short of eradication implies a continuing dedication of human and financial resources, together with perpetual surveillance and monitoring activities. A general overview of control methodologies for infectious dis-eases is given by Smith (1982), and a mathematical treatment of control by immuniza-tion by Anderson and May (1982), who consider the degree of reduction in disease incidence to be expected from various patterns and strategies of vaccination.

Where a formerly endemic geographic area has been cleared of transmission of a particular pathogen, the term *elimination* is often used. In such areas some means such as vigilance at borders may be employed to prevent or hinder reintro-duction, but the recurrent costs of surveillance must continue.

If transmission in the world is truly reduced to zero, the disease is *eradicated,* with no further need to spend resources for control or surveillance. Table 1–4 lists the basic features of diseases that may make them amenable to eradication. The International Task Force for Disease Eradication, formed in 1988, has evaluated the global eradicability of 21 infectious diseases and has determined that five are potential candidates for eradication. These are dracunculiasis, poliomyelitis, mumps, rubella, and taeniasis/cysticercosis. Unfortunately, partial achievement of this goal may bear a high price:

> Research on tuberculosis and BCG seems to have suffered from its own early success; that is, it seems as though the global BCG campaigns were more success-ful in eradicating tuberculosis research than in eradicating tuberculosis (Fine 1989).

The reduction of research on tuberculosis has been costly and shortsighted in view of the worldwide resurgence of this disease in the 1990s, and the appearance of many new multiply drug-resistant strains of *Mycobacterium tuberculosis.*

TABLE 1–4.    Factors Favoring Eradication of a Disease

---

- Infection and disease limited to human hosts, and transmitted person to person; no animal reservoir or insect vector
- Characteristic clinical illness, usually serious, and esily diagnosed
- Few or no subclinical cases
- No long-term carrier states
- Only one causative agent or serotype
- Short period of infectivity and after clinical illness
- Immunity following illness or immunization is of long duration, inhibits reinfection or reactivation, decreases or eliminates excretion of organism, and is detectable
- Disease has seasonality, permitting vaccine strategies during periods of low transmission
- Desirable characteristics of vaccine are safety, stability (resistant to physical and genetic change), possible transmission to others in the community, and absence of interference by maternal antibodies in infants
- Eradication would be cost-effective

---

*Source:* Adapted from Evans 1985.

Programs for immunization of infants and young children usually increase the mean age at first infection in the population and lengthen the period between disease outbreaks. As a rule, diseases producing a relatively small number of secondary cases in a disease-free population have a higher average age of attack, may be controllable by immunization, and may be candidates for eradication. It is generally best to immunize children as young as possible, for if the average age at which children are immunized is greater than the age at which they are normally infected, control is impossible and eradication out of the question. In the case of measles, current policy in developing countries calls for immunization with live virus vaccine at 9 months, and in the United States at 15 months, when seroconversion rates are highest. However, children much younger than the target age may be susceptible and become infected. Therefore the optimal vaccination policy must provide protection to young infants, who may still have circulating antibody transferred from their mother during fetal development. Since this maternal antibody neutralizes the live vaccine virus as well as "wild" disease-causing measles virus, it will be important to explore other means of immunization against measles.

Immunization is a tool for disease control mainly for pathogens transmitted person-to-person. Where there is a non-human reservoir, immunization of those animals could be useful if they are accessible, as in the case of rabies in household pets. Where human cases are derived primarily from disease cycles maintained in wildlife, as in sylvatic yellow fever or plague, immunization of people, while individually protective, will be of little value in reducing the risk of infection in others. Where the pathogen is ubiquitous in the environment, as with spores of tetanus in the soil, immunization of an individual will have no effect on the risk of infection of another, except for pregnant mothers and their infants. The proposed antigametocyte components of a future malaria vaccine would not help the vaccinee directly but could reduce transmission of the disease by aborting the infection in vector mosquitoes.

## Stages in the Genesis of a Product

The process of product development from basic research to routine deployment is shown for a hypothetical vaccine in Figure 1–1 and discussed in greater detail in subsequent chapters. The scientific understanding and technical support required to conduct the activities in each phase are shown, as well as a very brief summary of constraining variables. Reference is made also to the types of risk described more thoroughly later in this chapter. The process is not linear from one activity to another but includes feedback loops for modification and fine tuning at various points.

*Concept and development.* Basic research is concerned solely with the advancement of knowledge, and because one never knows the eventual outcome, it is not possible a priori to steer the results of basic research into a specific practical direction. The accumulated knowledge derived from prior basic and applied research forms the starting point for Figure 1–1, where a credible idea begins its long path to a potential product.

The relevance of basic biomedical understanding does not disappear with the start of directed vaccine development; indeed, there is a continuing need to fill in gaps in knowledge that become evident as the development programs progresses. Such studies may be called "postbasic research," meaning laboratory-based investigation that is not specifically applied, but is conducted with potential practical uses in mind. Much laboratory research falls into this category, particularly in today's world of the scientist-entrepreneur. For example, De Wilde (1987) has characterized the research phase for a yeast recombinant hepatitis B vaccine as: "(1) gene isolation, (2) gene characterization, (3) gene expression, (4) product isolation (analytical), (5) product characterization (biochemical and immunological)." Each of these procedures might be undertaken in the identical way as basic research in a university laboratory, except for the intent of producing a commercially viable vaccine.

*Product design and early clinical trials.* These activities, discussed in detail in Chapters 4 and 5, are fundamental because the vaccine product must be shown to

FIGURE 1–1. Diagrammatic representation of the process by which a new vaccine is derived and deployed. Common feedback loops among the various activities are indicated, as well as a rough indication of the type of science and technical support required, and of the limits and constraints as well as risks by type as listed in Table 1–6.

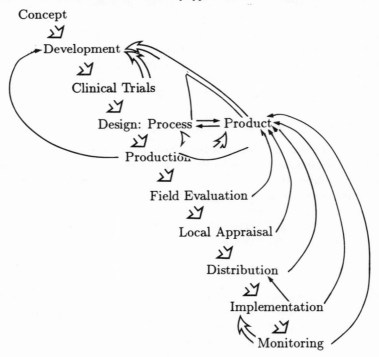

| Activity | Science | Technical Support | Limits and Constraints[1] | Risks |
|---|---|---|---|---|
| Concept | Basic | Library<br>Laboratory | Vision<br>Knowledge | Alternative<br>Perception |
| Development | Basic<br>Postbasic<br>Veterinary | Biotechnology<br>Laboratory<br>Animals | Knowledge/Technology<br>Biology<br>Validity | Alternative<br>Competitive<br>Feasibility |
| Clinical trials | Statistics<br>Pathology<br>Immunology<br>Clinical | Laboratory | Volunteers | Injury<br>Alternative<br>Competitive |
| Design<br>  Product<br>  Process | <br>Engineering<br>Engineering | <br>Laboratory<br>Laboratory<br>Pilot Plant | <br>Materials<br>Scale-up | <br>Competitive<br>Economic<br>Logistic |
| Production | Engineering | Plant | Materials<br>Process technology<br>Quality Control<br>Good Manufacturing<br>  Practices | Economic |
| Field evaluation | Epidemiology<br>Management<br>Social<br>Clinical | Health system<br>Information System<br>Diagnostic | Collaboration:<br>  Official/Public<br>Disease presence | Alternative<br>Ethical<br>Regulatory |
| Local appraisal | Political<br>O.R.[2]<br>Economics | | Health policy<br>Licensing | Cultural<br>Economic<br>Political<br>Regulatory |
| Distribution | Logistics<br>O.R.[2]<br>Management | Packaging<br>Cold chain | Infrastructure<br>Product stability<br>Legalities<br>Health facilities | Competitive<br>Cultural<br>Logistic<br>Regulatory |
| Deployment | Social<br>Management | Marketing<br>Mass media<br>Logistics | Literacy<br>Access<br>Acceptability<br>Availability<br>Administration<br>Motivation | Development<br>Economic<br>Environmental<br>Epidemiologic<br>Injury<br>Institutional<br>Opportunity<br>Political<br>Social<br>Sustainability |
| Monitoring | O.R.[2]<br>Epidemiology<br>Economics | Health System<br>Information System<br>Cost/Benefit | Records<br>Management<br>Information | |

[1] Money and management are presumed to be constraints at all stages.
[2] Operations Research

perform as expected to protect and not to harm; and it must be feasible to manufacture, store, and distribute. Design and testing go on at the same time, hence the feedback loops. In the "product design" step, for example, a particular antigen found in animal studies to be protective against subsequent challenge with a pathogen, has been produced in small quantities for laboratory use. Now the antigen must be combined with an adjuvant, carrier, preservative, or whatever is needed to make a complete product for eventual use in the field, and small scale human testing must be carried out.

*Scale-up and production.* Batches of the complete product can now be made. Scale-up procedures incorporating the process design step bridge the gap between laboratory-scale formulation of test batches and the methodologies by which production-scale quantities of the product can be manufactured, tested, and distributed. Quality control procedures must be incorporated to assure the purity, potency, stability, and uniformity of each lot. These procedures may not be implemented until the early evaluation phase shows promising results, but they must at least be considered fairly early. At the same time, account must be taken of statutes concerning pharmaceutical and biological products in both producing and receiving countries, as regulatory agencies become involved in the process of approval and registration of the product. The policies of the multilateral organizations that oversee immunization programs and distribute vaccines (primarily the World Health Organization and the United Nations Childrens Fund, respectively) must also be incorporated into the planning.

*Field evaluation and local appraisal.* As a general rule, it is easier to make a product than it is to find out whether it is useful. For example, it has been estimated that the initial scientific research on yeast recombinant-produced hepatitis B surface antigen vaccine accounted for only about 10% of the total resource cost of product development (De Wilde 1987). National and international regulatory agencies must approve the field trial protocols, monitor the evaluation as it proceeds, and determine whether the product is proven to be safe and efficacious. Individual health jurisdictions must decide whether to sanction and adopt the product based on its preventive efficacy, cost, and drawbacks, in comparison with competing products and modalities.

*Distribution and implementation.* During this phase the product becomes integrated into routine use in a "steady state" condition in the country involved. The primary actors are the manufacturer, the government (represented by its public health authorities), regulatory agencies, and the recipients, as well as those who pay for the product. These purchasers may be national or local governmental bodies, external donor agencies, the private medical sector, or individuals if the vaccine is on the open market. An ongoing program of monitoring or surveillance is needed, particularly in the early period of use, to identify rare adverse effects not found in clinical trials and to watch for case clusters indicating vaccine failure. In one example, an outbreak of paralytic polio in an immunized population in Brazil caused oral polio vaccine to be reformulated (Patriarca et al. 1988). In the most extreme case, influenza vaccines must be redesigned each season depending on the predicted prevalence of virus serotypes.

One cannot assume a priori that if there is (1) a genuine need, and (2) a proven relevant technology, the introduction of the technology will extinguish the need. For example, safe, efficacious, and relatively cheap measles vaccines have been available since 1963. Nevertheless, an estimated 1.5 million children a year still die from measles in nations (including the United States) that lack the resources, infrastructure, and motivation to vaccinate all of their children.

## Current and Emerging Needs for Innovative Biotechnology

*Prophylactic drugs and vaccines.* According to the World Health Organization's Special Programme for Research and Training in Tropical Diseases (TDR; see chapter 6), work was under way by June 1989 on the design of 70 to 100 different field trials of new drugs, vaccines, and vector control devices to combat malaria, leishmaniasis, trypanosomiasis, schistosomiasis, filariasis, and leprosy. Most trials were intended to begin by 1994, and some costly collaborative studies involving up to 100,000 people were expected to require 20 years or more to complete.

Considering vaccines alone, the total number of potential field trials may be far greater than those planned by the TDR program. Dozens of commercial companies and other organizations in many countries are striving to develop new vaccines (Edgington 1992). Assuming that many of these efforts will lead to potentially useful products, how and where are these to be tested, and what will be the consequences if and when they are implemented on a routine basis?

*Therapeutic vaccines.* It is possible, at least in theory, to make products that could be used for immunotherapy, or treatment of existing disease. Immunotherapeutic agents have been proposed for use against tumors, in cases of AIDS, and for other chronic and infectious diseases.

*Nondisease vaccines.* A major potential application of a nondisease vaccine is for the immunological control of fertility: a "birth control vaccine." The World Health Organization has had a Task Force on Vaccines for Fertility Regulation since 1973 to support basic and clinical research on this subject. Vaccines are under development for control of fertility in both males and females. A single-injection vaccine for the sterilization of males has been field tested. Potential candidate vaccines can also affect various stages of the fertilization process: some act against components of sperm to block human sperm-egg interactions. Circulating hormones essential for egg production and release may be neutralized by antibodies. Other targets occur later in the developmental process—for example, to prevent implantation of a fertilized egg. Phase I clinical trials have been carried out with anti-human chorionic gonadotropin vaccines in volunteers who had had tubal ligations (Jones et al. 1988; Talwar et al. 1990). Passive immunization with human monoclonal antibodies may provide a safe and effective method for contraception or termination of very early pregnancy.

Research in contraceptive vaccine development is inhibited by the political and economic pressures from groups opposed to such work. However, research is continuing. For example, a collaborative program in Contraceptive Development and Research in Immunology (CD&RI) has been under way for several years between the U.S. Agency for International Development and the Government of India. Community-based field trials of emergent products are well in the future and will require stringent planning and regulation.

*Potential and reality.* A note of caution was sounded by Edgington (1992), whose comprehensive study listed well over 100 proposed "biotech vaccines" but pointed out that only one (for hepatitis B) has actually come into widespread use: "Despite biotech's growing commitment to providing a vaccine for everything from AIDS to birth control, there remain serious obstacles—technical, regulatory, and legal—to their promise."

## Technology Assessment (TA) and Technology Forecasting (TF)

For thousands of millions of people in developing regions, the transfer, diffusion, and deployment of imported scientific innovation will engender certain consequences. The potential for improving health must be balanced against potentially high economic and other costs. The many perplexing uncertainties in this area call for objective analytic procedures to help support decisions for or against adoption of biomedical innovations.

Many authors have tried to define technology assessment. On utilitarian grounds, Brorsson and Wall (1985) considered that the "goal of TA is to provide decision-makers with valid and timely information concerning the general value of a given technology." Emphasizing process and a future orientation, Coates (1976) characterized TA as "a class of policy studies which systematically examine the effects on society that may occur when a technology is introduced, extended, or modified. It emphasizes those consequences that are unintended, indirect, or delayed."

Martuscelli and Faba (1993), of the National Autonomous University of Mexico, have anticipated much of the spirit and intent of the present volume:

The assessment of technology application in medical practice was basically concerned in its beginning with efficiency, intrinsical technical quality and the capacity to guarantee safety in the protection of life, or to avoid adverse effects or the death of the patient at a maximum. Besides efficiency, effectiveness and security analyses, with the presence of sophisticated technologies, nowadays costs, sensitivity and the specificity of the medical technologies which enter the market have also begun to be subjected to analyses. The social impact of technology in the population has become a core element in assessment studies. . . . The diversity of effects produced by the use of technology in medical care has turned assessment into a basic instrument to establish selection criteria and to facilitate their identification, incorporation and usage within the process of health care. Their projection at the level of society in general is attained when assessment results establish criteria and opinions to define policies and support those decisions which will influence procedures for the selection and incorporation of medical technologies.

Technology assessment is grounded in value judgments and policy issues. Technology forecasting differs in that attempts are made through extrapolation of trend lines, collection of expert opinion, or other means to project the extent of usage and the impact of a technology at some future time (Saren and Brownlie, 1983.) What is really being forecast is the environment in the future in which the innovation is expected to be used. There are many overlaps, but in general TA is more political; TF more commercial. As Bright (1978) has pointed out, the emergence of a technical capability may be forecast with some confidence, but forecasting the adoption and diffusion of that capability is far more complex and tenuous.

Although notoriously unreliable, technology forecasting is practiced because planners "have become increasingly aware of the need to anticipate technological change to reduce the risks and uncertainties associated with their decision-taking and planning activities" (Saren and Brownlie 1983). These authors state that "short term forecasts tend to be overly optimistic about what might be accomplished. Long term projections, covering several decades, tend to be too conservative in their estimations of what technology might achieve and of the social changes deriving from such achievements."

Managing the uncertainties of biomedical innovation in the developing world will require the tools of both TA and TF to help anticipate, insofar as we can foretell anything, the effects that are likely to occur when a particular biomedical technology is introduced into a specific developing country setting. The underlying purpose is to help assure, in an orderly manner, that the technology will be beneficial to the communities involved. We are interested especially in the early detection and projection of consequences, particularly those not anticipated by the designers and transmitters of the technology.

The Director-General of the World Health Organization has written:

> Haphazard approaches to technology development have led to imbalances such as neglect of services related to goods and less application to social development than to industrial development. Technical cooperation activities have not been sufficiently broad in scope to accommodate the whole range of changes needed if the full force of health technology is to be brought to bear on health problems. The difficulties of users in assimilating new technology have not been fully dealt with, and the roles of the parties to the mechanisms of technology transfer are not well established. . . . The relationship between provider and user countries must change from one-way dependence to interdependence among partners. . . . The process of transferring technology should include all the necessary information aspects, manpower development processes, and systems for proper application of the technology to the specific conditions of new users. When possible, components available locally should be used so that the transfer of technology may have more positive economic consequences. . . . As demonstrated in the case of microcomputers,

the appropriate technology may be an advanced rather than a simple one. Well-selected higher technology may be necessary for the adequate solution of some health problems. Health systems research is called for to assess the actual application of technology (Nakajima 1989).

## The Risk Audit

Survival or extinction of biomedical innovation will be determined by the weakest link in the long pathway of technology transfer and diffusion (Figure 1-1) from the originator to the receiver—the person for whose benefit the procedure is intended.

### A Look at Benefits

As with most important terms, the precise meaning of "benefit" is elusive, and it need not be argued whether a benefit "should be defined as what people like or what paternalistic others believe to be good for them" (Fischoff and Cox 1986). In general, Trojan horses aside, people can recognize a benefit without exhaustive analysis.

Benefits of an introduced technology must be viewed in terms of the goals of the receiving individuals and of their medical care system and society in general. An enhanced ability to diagnose a disease may result in earlier treatment, a shorter period of suffering by the patient, perhaps an easier and less costly course of treatment, and reduced risk of more severe illness or perhaps death. At the same time the health system may save resources by avoiding additional visits to the health center, further costs incurred by a patient with more advanced disease, and so forth. Benefits to society may ensue, for example, in the form of increased equity and access to medical care through saving of resources, thereby making facilities and staff time available for other patients. The additional economic productivity gained from the healthy person, or from the mother of the no longer ill child, can also be significant. In a non-disease example, family planning programs have become popular because both the couples that use the methods ("acceptors") and society in general are intended beneficiaries.

A continuing problem in cost-benefit analyses is the inability of quantitative methods to deal with the nonquantitative issues of quality of life, pain, family stress, and so on, that accompany illness and disability. Another problem is the trade-offs among different uses for resources and the different kinds of benefits obtained. There is no question that immunization against major diseases will reduce deaths from those causes among infants and children. But consider the observation of Rahmathullah et al. (1990) that young children in south India who received small weekly doses of vitamins A and E had less than half the mortality rate of those receiving vitamin E alone. How can one compare the beneficial effects of vitamin A supplementation at $2 per child per year with the estimated $10 to $15 cost of a fully (and specifically) immunized child?

The cost of a technology introduced into a developing country is a major element in its transferability, but the absolute cost in dollars or rupees or pesos is not as important as affordability within a specific context. A price considered a bargain by one person may be prohibitive to another. The issue of costs and benefits of technological innovations is discussed further in chapter 5.

Despite its pervasive significance, the model of biomedical technology transfer presented here is grounded not on benefit but on risk because (1) in general the benefit of any innovation should be self-evident; if it is not, it is unlikely to be adopted in any event; (2) the risks may be more broadly distributed and more difficult to perceive; and (3) people are more likely to act to avoid real (or perceived) risks than to secure potential benefits.

## A Taxonomy of Risks

There are always costs and risks inherent in the international transfer and deployment of such products. Recognition of the questions and uncertainties associated with the international transfer of biomedical innovations is a first step in making the process more efficient and less costly. However, few laboratory researchers are knowledgeable about implementation, particularly in developing countries, and local public officials who understand the technical aspects of modern biomedicine may be fewer still. Other parties with an interest in this kind of technology transfer include decision makers in national governments and intergovernmental organizations; businessmen, who may play an essential role in production, distribution, and marketing of the innovations; members of the medical and health professions; and the residents of the developing countries who are the participants, subjects, and targets of the transfer of technology.

All aspects of life are pervaded by the concept of risk, which acknowledges our inability to control fully the consequences of our actions and considers the likelihood of occurrence of certain alternative future events, some of which may be adverse to our interests. The more precisely their nature and magnitude of relevant risks can be characterized, the more informed and appropriate will be the decision making. Table 1–5 shows twenty partially overlapping categories of risks that may be incurred in the transfer and deployment of biomedical innovations, and Table 1–6 indicates in very general terms some steps that may be taken to mitigate these risks. A discussion of these steps occupies much of the remainder of this volume.

## Acceptability of Risks

When in the past a new product, industry or agent was introduced into widespread use, there was little effort or felt need to predict or anticipate possible consequences to health that might be associated with that innovation. One wonders how the automobile might have been accepted if it had been known that its use would be at the price of 50,000 lives per year in the United States alone. . . . In an affluent society such as ours, the benefits of technology are no longer so irresistible that we are willing to overlook possible costs, particularly when those costs are to health and when such costs are likely to be subtle and overlooked unless assiduously examined (Sagan 1984).

TABLE 1–5.  Twenty Types of Risk in Transfer of Biotechnologic Innovations (BI) to Developing Countries

| Type of Risk | Example of Risk |
| --- | --- |
| Alternative | A different path can achieve a similar outcome |
| Attainment | BI is so successful it is no longer needed |
| Competitive | Equal or better BI product, marketing, or acceptability |
| Cultural | People will reject or misuse the BI |
| Development | Receiving country cannot properly exploit the BI |
| Economic | Monetary loss, excessive cost, or wasted time or effort |
| Environmental | Damage to animals, plants, physical environment |
| Epidemiologic | BI is ineffective or the health problem is made worse |
| Ethical | Testing or use of BI contravenes moral or ethical precept |
| Feasibility | What is intended cannot be done for some reason |
| Injury | BI will have adverse effects on health of receivers |
| Institutional | BI may be threat to status quo of certain interest groups |
| Logistic | Production, supply, transport, storage, or administrative problem |
| Opportunity | With zero-sum resources, if this BI is introduced, what else must be curtailed? |
| Perception | The presumed problem is not what it appears to be |
| Political | Unfavorable political repercussion from use of the BI |
| Regulatory | Legal, patent, or regulatory barrier to implementation |
| Social | Some social dislocation may result from use of the BI |
| Strategic | The planned strategy will not achieve the desired goal |
| Sustainability | BI will disappear with loss of external support |

*Societal, occupational, and individual acceptance.*  The degree of acceptance of, or aversion to, particular risks is an expression of an individual's personality, cultural ties, and unique situations with relation to the perceived benefit and risk involved, and of his or her ability to identify and evaluate the risk. In the case of the automobile just mentioned, most people seem willing to accept the risks involved as a fair exchange for the desired benefits; somewhat fewer are willing to ride in airplanes, and fewer still seem anxious to try hand gliding or sport parachuting.

Studies by Slovic et al. (1980) show that the risks of low-frequency events, such as fatalities from homicide, botulism, fires, snakebite, tornados, and abortion, tend to be overestimated. Conversely, risks of high-frequency events such as deaths from asthma, stroke, and diabetes, tend to be underestimated.

In one of the few studies of comparative societal views of risk, Keown (1989) compared the perceived degree of risk from certain substances and activities between U.S. citizens and Chinese residents of Hong Kong. In general, Hong Kong Chinese thought that food coloring, heroin, and caffeine carried substantially more risk, while DNA research, pesticides, and oral contraceptives had considerably less risk, than did the Americans. Such views will affect the degree of acceptance of certain imported technologies and may determine local success or failure.

TABLE 1–6.  Appropriate Responses to Perceived Risks in Transfer of Biotechnologic Innovations (BI) to Developing Countries

| Type of Risk | Steps to Minimize Risk |
| --- | --- |
| Alternative | Improve relative effectiveness and efficiency of BI |
| Attainment | Bravo! |
| Competitive | Awareness, improve system efficiency, cost reduction |
| Development | Adapt the BI for ease of use; bolster infrastructure |
| Cultural | KAP surveys, observation, education, interaction |
| Economic | Improve efficiency of resource utilization |
| Environmental | Survey and monitor biota and environment |
| Epidemiologic | Followup observation, diagnosis, monitoring |
| Ethical | Follow guidelines, organizational, and review procedures |
| Feasibility | Assure that the intended result is achieveable |
| Injury | Better preliminary studies; monitor pathogen and host; alter BI |
| Institutional | Liaison with groups affected; reduce perceived threat |
| Logistic | Simplify scaleup and production; quality control; assure systems for supply, storage, and transport; improve administration and procedures; educate users |
| Opportunity | Establish priorities |
| Perception | Check statistics, diagnoses, etc. to verify the problem |
| Political | Attention to all risks; observe local regulations and customs |
| Regulatory | Know legal and judicial systems; obey rules and regulations |
| Social | Use BI in nonthreatening manner |
| Strategic | Understand the system; do sufficient preliminary studies |
| Sustainability | Develop appropriate systems to maintain support |

A single technology, such as an immunization, may be viewed very differently by different groups of people. Some will wait for hours or storm the hospital gates in an effort to obtain a highly valued "jab" during a cholera epidemic in Southeast Asia, whereas others, with sincere conviction, will prayerfully accept fines and prison sentences to prevent their children from being immunized against measles in Philadelphia.

Businessmen and investors are, to a large part, professional risk-takers; scientists and politicians less so; and ordinary people in developing countries, as a rule, avoid taking risks. Each actor in the transfer process, from the originator to the receiver, will have a particular relationship to the innovation, and, from his or her viewpoint, each encounters a unique set of potential risks in the designated categories (Table 1–7), as well as a complementary set of potential benefits. In relation to any specific innovation, each individual, deliberately or subconsciously, will

1. Identify some potential benefits and risks, valid or fanciful
2. Estimate their magnitude

3. Evaluate their personal, familial, or social significance
4. Consider ways first to minimize risk, then to increase benefits for him or herself

*Balancing among risks.* Biomedical science is continually changing, and so are actual and perceived risks of biomedical technology. It may be reasonable to implement an innovation, such as a vaccine, in one geographic context and not in another; an innovation worth using today may be rejected tomorrow if the epide-

TABLE 1–7.  Generic Taxonomy of Types of Risks and Levels of Concern, by Realm, in Transfer of Biomedical Innovation to Developing Countries

| Type of Risk | Level of Concern, by Realm | | | | | |
|---|---|---|---|---|---|---|
| | Originator | Evaluator | Producer | Implementor | Performer | Receiver |
| Alternative | H | L | H | H | L | L |
| Attainment | L | O | M | L | L | O |
| Competitive | M | O | H | O | O | O |
| Cultural | L | H | M | H | H | H |
| Development | L | L | H | H | M | O |
| Economic | M | M | H | H | O | H |
| Environmental | L | M | L | M | M | L |
| Epidemiologic | H | H | H | H | H | H |
| Ethical | M | H | M | H | M | M |
| Feasibility | H | H | H | H | M | M |
| Injury | H | H | H | H | H | H |
| Institutional | O | L | M | M | L | L |
| Logistic | L | H | H | H | H | O |
| Opportunity | L | H | M | H | L | H |
| Perception | M | H | H | M | L | L |
| Political | L | L | H | H | M | M |
| Regulatory | M | M | H | H | L | O |
| Social | O | M | M | M | L | H |
| Strategic | H | H | H | M | M | M |
| Sustainability | L | L | H | M | O | L |

Level of Concern: H, High; M, Moderate; L, Low; O, Little or None

Originator: The bench scientist or laboratory research team that invented or produced the innovation

Evaluator: The consortium of interests that performs laboratory and field testing and assessment of the innovation

Producer: The financial and commerical interests that manufacture and market the innovation

Implementor: The receiving government agencies and others concerned with routine deployment of the innovation in the community

Performer: The individual, usually a trained health worker, who performs the procedure or applies the innovation to or for the receiver

Receiver: Members of the community or population in which the innovation will be implemented in practice

miologic, technologic, economic, or regulatory environment changes. An attempt to reduce one type of risk may increase another. For example, Moore (1983) cites the banning of DDT by Sri Lanka on environmental grounds, thereby preventing its use during a subsequent malaria outbreak in which many people died.

## Having an Impact

In today's instantaneous world of cellular telephones and on-line data bases, we have come to expect rapid results from our investments of ingenuity, effort, and resources. Yet, in the struggle against infectious diseases, the most potent ally may be patience. The biomedical innovator whose product can survive the risks may hope to find his or her perseverance rewarded. Of course, it may take some time for the innovation to have an impact in the form of improved human health. Edward Jenner's predicted conquest of smallpox has, ultimately, been witnessed by his great-great-great-great-great grandchildren.

# 2

# Health in Developing Countries: The Role of Vaccines

## Economic Development

Of the five billion people currently alive, about 75% are poor or very poor by western standards. Within most developing countries the proportion of impoverished people is even greater, and the socioeconomic distance between the wealthier and the poorer elements is larger than is found in the industrial countries.

No universally accepted definition of a "developing country" exists. Few would argue about Bangladesh, Bhutan, Chad, or Bolivia. The larger, more diverse countries such as Mexico and Brazil have many elements of developed areas, particularly in their science and technology sectors. At the same time, some European countries such as Portugal, Albania, and Bulgaria share many characteristics with the so-called Third World. Other countries are undergoing negative development and becoming poorer.

It is commonly held that nations follow a linear progression from undeveloped through developing to developed, and that examples can be found to illustrate any stage of development. In fact, each country and region proceeds along its unique path depending on local political and economic systems, concentration of business ownership, agrarian organization, the competence of planning, and openness to outsiders.

In the 1950s, definitions of development were based on industrialization, considered to be the means by which people in poor countries would acquire the wealth to adopt the consumption patterns of people in rich countries. Therefore investment was centered around factories, power plants, dams, roads and bridges, and agricultural projects. Subsequently, development economists realized that industrialization, urbanization, and modernization were impossible without concurrent investments in "human capital" through education, health, housing, and other social programs. Definitions of development began to include advancements in "equity" and "socioeconomic status," which refers not only to the amount of money or resources at the command of individuals, but to their perception and acceptance of modernization, and their power to influence the system.

## Variations within Developing Countries

Despite the superficial simplicity of single-number representations of countrywide data (e.g., per capita GNP is $1,211; infant mortality rate is 79 per thousand live births), the fact is that all countries are socioeconomic chimeras. Circumstances within individual countries are fully as diverse as those among them. The range of income levels is broad, and their distribution is unequal based on ethnic group, urban and rural residence, educational attainment, occupation, and so forth. Intra-country differences are pronounced, with mixed demographic and epidemiologic patterns characteristic both of prototypical poor countries (high birth and death rates, predominance of infectious and parasitic diseases, numerous children and few elderly) and industrialized countries (low birth and death rates, predominance of chronic diseases, relatively fewer children and more elderly). Most countries include islands of wealth in a sea of poverty, or vice versa, with different health challenges and remedies for each situation.

Impoverished people in tropical developing countries have many characteristics in common, almost independent of their nationality or ethnic group. Conditions vary, but as a rule:

1. About four-fifths of family income is spent for food. The diet is monotonous, and commonly limited to cereals, yams, or cassavas, a few vegetables, and in certain regions, a little fish or meat.
2. Undernutrition is common and many are severely malnourished. Energy and motivation are reduced; performance in school and at work (if there is any) is undermined; resistance to illness is low; and the physical and mental development of children is often impaired.
3. Birth rates are high, as are mortality rates for infants and children under five. The expectation of life at birth is low, and about 50% attain the age of 40.
4. Vaccination rates are low; immunizable diseases are widespread and, as in the case of measles, are more likely to be fatal in poor areas than in more prosperous communities within their own country, and far more than in the wealthier countries.
5. Only about one-third of the adults are literate.
6. Roughly 20% of children complete more than 3 years of primary school.
7. The status of girls and women is low, and early marriage is common. (Based on and adapted from Nafziger 1984.)

Even more than cash income, the state of literacy and knowledge, adaptive social and cultural attitudes, and a progressive outlook determine receptivity to importing and establishing technology. The privileged class has access to the benefits of modernity and is generally receptive to incorporating change into everyday life. Not just intended receivers or targets of an innovation, this group participates actively in the process of technology transfer.

## Differences among Developing Countries

The annual compendia of the World Bank *(Atlas, World Development Report)*, United Nations Development Program *(Human Development Report)*, UNICEF *(State of the World's Children)*, the World Health Organization *(World Health Statistics Annual)*, and other agencies provide dozens of charts and tables categorizing and grouping countries and summarizing national characteristics. Students unfamiliar with global finance are often surprised at the economic debility of many developing country governments, when compared to large transnational corporations. Table 2–1 shows changes in the gross national product (GNP) of a number of developing countries, as well as financial turnover and growth in sales of a score of major pharmaceutical companies, to illustrate their relative clout.

## The Debt Crisis and Health in Developing Countries

Since the mid-1970s, recurrent economic crises have caused a steady degradation of medical care and social programs in many developing countries. The economic deterioration has occurred after large amounts of money were borrowed from international lenders. Loans, obtained from official agencies and private banks replete with money from oil-rich nations, were intended to help pay for development programs. Repayment was expected from profits generated by economic development. As it happened, high interest rates, reduced international aid flows, falling internal and external investment, and other factors have interacted so that many developing countries are unable to repay their loans. Some have made matters worse by borrowing more money just to pay the interest.

At the end of the 1980s the debt owed by the developing world to the wealthier countries exceeded US$1000 billion. Debt repayments, on average, accounted for about one-fourth of export revenues to developing countries. Approximately US$20 billion per year flowed from the Third World to the wealthier countries. At the same time, prices for primary products produced by the poorer countries had fallen by almost one-third throughout the 1980s.

The health effects of the debt crisis in the poorest countries may be viewed in the following context:

1. The effective per capita income of many hundreds of millions of people is declining. On average, the residents of Latin America experienced a 9% reduction in income during the decade of the 1980s. In sub–Saharan Africa the situation is worse. "Africa enters the 1990s embattled, laden with debts, and effectively recolonized" (Alubo 1990).

2. The decline in personal income and increase in unemployment means that minimum- and low-wage earners and their dependents experience further deterioration in nutritional and health status.

3. Decreased discretionary income will cause individuals to forgo private fee-for-service health care and be forced to utilize free or nearly free government health services of lesser quality.

4. Lowered employment leads to reduced tax collections and less revenue for social security and other government programs.

5. Structural adjustment policies are ordered by the International Monetary Fund to

TABLE 2–1.  Comparison of Growth of Sales of 20 Leading Pharmaceutical Companies with the Growth in GNP of 25 Countries Whose GNP is Close to the Turnover of the Companies (1983–1986) ($ million)

| Company or Country | 1983 | 1984 | 1985 | 1986 | Average Yearly Change (%) |
|---|---|---|---|---|---|
| Costa Rica | 2,550.0 | 2,975.0 | 3,380.0 | 3,996.0 | 13.9 |
| Bolivia | 3,060.0 | 3,348.0 | 3,008.0 | 3,900.0 | 6.7 |
| Yemen | 3,410.0 | 3,575.0 | 3,740.0 | 3,850.0 | 4.0 |
| Uganda | 3,212.0 | 3,450.0 | 3,565.0 | 3,680.0 | 4.4 |
| **Merck Sharp & Dohme (USA)** | **2,422.0** | **2,656.6** | **2,824.0** | **3,441.0** | **10.9** |
| Honduras | 2,747.0 | 2,940.0 | 3,168.0 | 3,330.0 | 6.2 |
| Cyprus | 2,576.0 | 2,555.0 | 2,653.0 | 3,052.0 | 5.3 |
| **Hoechst (Germany)** | **2,552.7** | **2,295.4** | **2,396.4** | **3,042.6** | **4.7** |
| Mozambique | 3,059.0 | 3,151.0 | 2,240.0 | 3,003.0 | −4.1 |
| Afghanistan (estimated 1984–1986) | 2,414.0 | 2,601.0 | 2,805.0 | 2,924.0 | 6.2 |
| Lebanon (estimated (1984–1986) | 2,782.0 | 2,782.0 | 2,889.0 | 2,889.0 | 1.2 |
| **Ciba-Geigy (Switzerland)** | **2,108.7** | **2.059.3** | **2,277.6** | **2,851.2** | **9.1** |
| **Bayer (Germany)** | **2,430.4** | **2,128.9** | **2,267.3** | **2,787.5** | **3.5** |
| Senegal | 2,728.0 | 2,394.0 | 2,368.0 | 2,772.0 | −0.2 |
| Nicaragua | 2,728.0 | 2,666.0 | 2,464.0 | 2,686.0 | −0.8 |
| Iceland | 2,052.0 | 2,204.0 | 2,143.8 | 2,682.0 | 8.1 |
| Albania (estimated 1984–1986) | 2,436.0 | 2,520.0 | 2,604.0 | 2,604.0 | 2.2 |
| Papua New Guinea | 2,660.0 | 2,485.0 | 2,380.0 | 2,592.0 | −1.1 |
| **American Home Products (USA)** | **2,333.2** | **2,416.7** | **2,523.5** | **2,560.4** | **3.0** |
| Nepal | 2,512.0 | 2,576.0 | 2,640.0 | 2,535.0 | 0.3 |
| Madagascar | 2,945.0 | 2,522.0 | 2,400.0 | 2,369.0 | −7.7 |
| Haiti | 1,890.0 | 2,080.0 | 2,046.0 | 2,244.0 | 5.4 |
| **Pfizer (USA)** | **1,866.0** | **1,891.0** | **1,961.0** | **2,203.0** | **5.3** |
| **Sandoz (Switzerland)** | **1,450.4** | **1,405.3** | **1,592.2** | **2,155.1** | **11.5** |
| **Glaxo (UK)** | **n.a.** | **1,400.2** | **2,118.1** | **2,143.2** | **17.5** |
| **Eli Lilly (USA)** | **1,645.8** | **1,664.5** | **1,786.0** | **2,119.8** | **7.9** |
| **Roche (Switzerland)** | **1,497.0** | **1,411.0** | **1,546.5** | **2,115.0** | **9.8** |
| Zambia | 3,596.0 | 3,055.0 | 2,613.0 | 2,070.0 | −20.3 |
| **Abbott (USA)** | **1,599.0** | **1,705.7** | **1,866.0** | **2,057.0** | **8.0** |
| **Warner Lambert (USA)** | **1,405.0** | **1,717.4** | **1,872.0** | **2,041.0** | **11.6** |
| Guinea (estimated 1986) | 1,560.0 | 1,881.0 | 1,952.0 | 1,984.0 | 7.4 |
| **Bristol-Myers (USA)** | **1,505.0** | **1,586.6** | **1,753.0** | **1,961.7** | **8.4** |
| Jamaica | 2,990.0 | 2,645.0 | 2,162.0 | 1,932.0 | −15.8 |
| **SmithKline (USA)** | **1,463.7** | **1,541.6** | **1,654.1** | **1,896.0** | **8.2** |
| **Upjohn (USA)** | **1,326.0** | **1,487.4** | **1,593.0** | **1,863.0** | **10.7** |
| Rwanda | 1,539.0 | 1,652.0 | 1,708.0 | 1,827.0 | 5.5 |

31

continued

TABLE 2–1.  Comparison of Growth of Sales of 20 Leading Pharmaceutical Companies with the Growth in GNP of 25 Countries Whose GNP is Close to the Turnover of the Companies (1983–1986) ($ million) *(continued)*

| Company or Country | 1983 | 1984 | 1985 | 1986 | Average Yearly Change (%) |
|---|---|---|---|---|---|
| Congo | 2,091.0 | 1,938.0 | 1,887.0 | 1,782.0 | −5.5 |
| **Johnson & Johnson (USA)** | **1,175.8** | **1,319.7** | **1,439.8** | **1,731.7** | **12.0** |
| **Takeda (Jpn)** | **1,291.0** | **1,297.0** | **2,226.4** | **1,700.0** | **3.8** |
| **Wellcome (UK)** | **n.a.** | **1,240.6** | **1,235.9** | **1,675.7** | **12.9** |
| Niger | 1,392.0 | 1,121.0 | 1,525.0 | 1,638.0 | 3.1 |
| **Boehringer Ingelheim (FRG)** | **1,238.7** | **1,259.0** | **1,309.4** | **1,616.9** | **8.2** |
| Mongolia (estimated 1984–1986) | 1,404.0 | 1,443.0 | 1,482.0 | 1,560.0 | 3.4 |
| **Schering-Plough (USA)** | **n.a.** | **1,199.7** | **1,274.0** | **1,557.6** | **12.0** |
| Barbados | 1,215.0 | 1,311.0 | 1,389.0 | 1,545.0 | 7.7 |
| Mali | 1,216.0 | 1,099.0 | 1,215.0 | 1,494.0 | 5.9 |

*Source:* Chetley 1990.

Note: n.a. = not available

help assure that interest payments continue to flow to foreign creditors. Austerity measures may include currency devaluation, the cancellation of subsidies for common consumer products, and cuts in public spending for social services, including education, welfare, and health. Real expenditures on health declined by as much as 50% in many poor countries during the 1980s (Grant 1989).

6. Some developing countries are experiencing inflation rates that are often hundreds, sometimes thousands, of percent annually. Inflation and devaluation lower the value of local currency and increase the cost of imports. Health budgets, already under pressure, can now afford fewer pharmaceuticals, medical supplies, computers, vehicles, gasoline, and all other imported items needed by the health service.

7. The double-edged pressure of increased demand and reduced capacity results in degradation of governmental health programs, expressed as decreased output of services, deterioration in quality, or both.

8. To conserve the output of service units to the public as much as possible, and because personnel commitments are likely to be honored (even at reduced salaries, if necessary), government health programs will preferentially cut or eliminate maintenance, new construction, and other forms of investment. Such strictures may help to maintain services in the short run but eventually will be counterproductive.

These difficulties, particularly the adverse effects of structural adjustment policies, have caused several countries to establish specific programs to mitigate SAP-caused hardships. A program known as "Adjustment with a Human Face" has been promoted by UNICEF and other agencies, and described in a book of the same name (Cornia et al. 1987).

Under these circumstances it is reasonable to presume that any technologic innovation that improves the cost-effectiveness of health interventions would be welcomed by national authorities. Still, no guarantee exists that funds will be available to subsidize its widespread adoption. Lack of resources in the recipient country can hinder the distribution and administration of pharmaceuticals and vaccines, even when these are obtained free of charge from external agencies. As an example, Merck and Co. has offered to donate millions of doses of ivermectin, a drug active against onchocerciasis (river blindness) in West Africa, but as Tanouye (1992) pointed out, "giving a drug away proved to be tougher than selling one." After five years "the drug has reached about 5% of the more than 100 million people who either have, or are at risk for, the disease. An inadequate health care delivery infrastructure, coups d'état, and rough terrain have presented big roadblocks to the giveaway. . . . Although Merck is giving the drug free of charge, some countries still can't afford to deliver it to those in need."

According to da Silva (1983), assistance from donor countries for biotechnology projects in developing countries is governed by four criteria:

1. The project should address the technical, social, and economic needs of the recipient country.
2. The project should involve shared responsibilities by both partners in the development and management of the project when and where necessary.
3. The project should be aimed at the promotion of self-reliance in the developing countries.
4. The final result of the project should be of benefit to both partners.

## Official Development Assistance

The "foreign aid" business amounts to about US$36 to $40 billion per year in bilateral flows from wealthy to poor countries. Most industrialized countries maintain bureaucracies devoted to development assistance to poorer countries. The United States and Japan compete as the largest donor, depending on the relative value of the yen and the dollar. A minor portion of these development funds is expended on the health sector, and much less is devoted to the transfer of biomedical technology.

Among the many critics of international development assistance programs is Bertrand (1989) who points out:

> Such systems of aid tend, first of all, to create relationships of political dependence between donors and recipients. . . . The fragmentation and overlapping of bilateral assistance organizations (which may number as many as twenty or so, with one, two, three or four being predominant, depending on the country) create difficult problems for the recipient countries in coordinating overall external aid and in conducting project-by-project negotiations. The "tied" nature of such aid results in disadvantageous economic conditions due to the obligation to buy from donor countries.

Hancock (1989) is more vituperative. He sees nothing but greed, waste, and incompetence in the international disaster and development assistance efforts of people he considers "well meaning but ignorant humanitarians."

## Science and Technology, Health, and Economic Development

Some consider that targeted programs of "Essential National Health Research" are necessary to reduce disease, improve survival, create a healthy population, and facilitate economic development (Commission on Health Research for Development 1990; Task Force on Health Research for Development 1991). Many believe that specifically directed biomedical research in developing countries, conducted by northern industrialized country scientists without regard to local health and medical care systems, is harmful (Banerji 1990).

An overriding question is the extent to which any biomedical technology can alter the conditions that foster the spread of illness and premature death. It may be that improvements in the amenities of life and in economic conditions will greatly reduce the diseases of poverty. For example, Esrey et al. (1991) analyzed 144 published studies to determine the health impact of water and sanitation interventions on the prevalence and intensity of a number of diseases. For the more rigorous studies, median reductions in morbidity were dracunculiasis, 78%; schistosomiasis, 77%; ascariasis, 29%; trachoma, 27%; diarrhea, 26%; hookworm, 4%. For hookworm infection, ascariasis, and schistosomiasis, there was a great reduction also in intensity of infection as determined by reduced worm egg counts. "For the rigorous studies the median reduction in diarrhoea-specific mortality was 65% and in overall child mortality, 55%, which suggests the important role that water and sanitation play in enhancing child survival" (Esrey et al. 1991).

Considering the potential effectiveness of a biotechnologic intervention in reducing the burden of these diseases, water supply and sanitation represent an alternative way to achieve a similar result. The effects of vitamin A supplementation alone in reducing early child mortality are similarly convincing.

> It appears plausible to assume that a certain take off in development is a prerequisite to rapidly falling mortality. A government structure conducive to socioeconomic development, progress in education, road communications, and an administrative infrastructure, even rudimentary, seem to play an important role in the initial stage. Once the process of decline is under way, public health measures, supported and coordinated in many cases at international level, become increasingly important and the decline takes place with such rapidity that improvements in levels of living are outdistanced. However, it seems that an expectancy of life at birth of 55–60 years constitutes a point beyond which the social, political, and economic factors again become increasingly operative [World Health Organization (WHO) 1974].

If this pattern is generally applicable, then health measures such as immunization should be most effective in reducing mortality in the middle stage of development. Most people in developing countries today are in this stage.

Because of lowered mortality, some critics have accused public health measures of responsibility for increases in world population and widespread environmental degradation, a view discussed at length in Chapter 4.

*Evaluating the threats to health.* The burden of illness in a population is estimated by various methods. Morbidity rates are based on properly diagnosed individuals whose records are accurate, adequately maintained, and forwarded through
the system. Often one or more of these elements is lacking so that reliable figures
are unavailable. For example, Ashford et al. (1992) surveyed published estimates
of the number of people infected with leishmaniasis, and found that authorities
differed by as much as 45-fold. A figure of 12 million annual cases, based
on WHO sources and "amplified by unspecified reports," has been repeated in a
number of publications, while several authors have used an estimate of 1.2 million
(which may have originated as a decimal error). Using the best population figures
and epidemiological data available, Ashford et al. systematically estimated the
annual incidence of visceral leishmaniasis at 88,500 cases and for cutaneous leishmaniasis at 295,900 cases, yielding a combined total under 400,000 cases.

When individuals devote great time and effort, or perhaps an entire career, to
research or control of a particular disease, it is understandable that they emphasize
the importance of "their" subject. Distortion of reality, however innocently derived, works against the scientists and administrators who must have a clear vision
of the epidemiologic situation. The significance of a vaccine differs substantially
if the true annual incidence is 12 million or if it is 400,000. The enthusiasm with
which effort and resources are invested in research, product development, and
evaluation is based on confidence in the validity of reported figures.

Even if prevalence estimates are grossly mistaken, there are certain generally
acknowledged hazards of life in poor tropical communities related to malnutrition,
lack of sanitation and education, and similar factors. To these ubiquitous illnesses
(mainly diarrheal and respiratory) are added Chagas' disease, onchocerciasis, Lassa
fever, and others of local significance. Commonly used estimates of the impact of
these diseases are the years of potential life lost (YPLL) owing to each cause or,
conversely, quality adjusted years of life gained (QALY) from various interventions. Causes of death that typically strike young people, such as malaria and
diarrheal or respiratory diseases, will naturally be weighted far differently from
the chronic diseases characteristic of the elderly.

*Causes of death.* In many countries deaths, particularly of young children, are
poorly recorded and cause-of-death data are unreliable. In Thailand, Lumbignanon
et al. (1990) carefully recorded all births and infant deaths in a rural district of
80,000 inhabitants over one year and compared their findings with official figures.
The authors' survey found 1,258 live births (official number, 832), 28 perinatal
deaths (0), 16 neonatal deaths (2), 29 infant deaths (14) and an infant mortality
rate of 23.1 per thousand live births, compared with the officially stated rate of
16.8.

It is just where disease occurrence is highest that the numbers are the least
trustworthy. The more preventive measures are needed, the harder it may be to
substantiate their level of effectiveness. For logistic reasons, studies may tend to
be done where they can be carried out more quickly, easily, and cheaply, and not
necessarily where conditions in the field will test the limits of the innovation.

Populations everywhere are aging. In the United States in 1970 the median age was below 28, in 1989 it was 33, and it is expected to rise to more than 36 by the year 2000. Causes of death in the advanced developing countries are changing in line with the demographic and epidemiologic transitions (Mosley et al. 1990).

Frenk et al. (1989) have compared changes in cause-of-death statistics in Mexico between 1955–1957 and 1980 (Table 2–2). Not only has the ratio of causes changed, but the absolute number of deaths per 100,000 inhabitants has declined sharply. Causes of death in developing countries can also vary greatly between urban and rural areas.

*The case of AIDS.* The rate of new infections with human immunodeficiency virus (HIV) is declining in the United States and other industrialized countries, but rising sharply in parts of the developing world. Officials of the Global Program on AIDS of the World Health Organization estimated in 1993 that more than 14 million adults were infected in 162 countries, with an expectation of approximately 40 million infections by the year 2000. The number of clinical cases was about 2 million by mid-1993, and expected to rise to about 10 million in the year 2000, 90% in developing countries, leaving approximately 10 million children orphaned.

A computer model of demographic indicators of a "typical" sub–Saharan African country had projected a decline in crude mortality rate from 14.8 per thousand inhabitants in 1985 to 8.4 per thousand in 2010 in the absence of AIDS, but predicted an increase to 16.4 per thousand if AIDS prevalence rates continue to grow as expected. Life expectancy at birth, which was earlier projected to rise

TABLE 2–2.   The First Ten Causes of Death, Mexico, 1955–1957 and 1980

| 1955–1957 | | 1980 | |
|---|---|---|---|
| *Cause of Death* | *Rate/100,000* | *Cause of Death* | *Rate/100,000* |
| Gastroenteritis | 227.5 | Heart Diseases | 74.9 |
| Influenza and Pneumonia | 202.0 | Accidents | 71.1 |
| Childhood diseases | 135.3 | Influenza and Pneumonia | 56.9 |
| Heart diseases | 91.4 | Enteritis and other diarrheal diseases | 55.1 |
| Malaria | 66.4 | Malignant tumors | 39.2 |
| Accidents | 48.1 | Certain causes of perinatal mortality | 39.2 |
| Homicide | 38.0 | Cerebrovascular diseases | 22.6 |
| Malignant tumors | 37.8 | Cirrhosis and other chronic liver diseases | 22.1 |
| Bronchitis | 31.7 | Diabetes mellitus | 21.7 |
| Tuberculosis, all forms | 31.2 | Nephritis and nephrosis | 10.5 |
| All other causes | 390.0 | All other causes | 231.6 |
| Total | 1299.4 | Total | 644.9 |

*Source:* Frenk et al. 1989.

about 10 years (from 51.5 to 61.4), may instead decline by four years (to 47) because of the effects of AIDS (Gladwell 1991). Such projections are based on current knowledge and technologies; a vaccine to prevent AIDS (see chapter 6) would change these figures dramatically.

## Child Health in the Developing Countries

In 1990 more than 4 billion people, or over 80% of the world's population, lived in developing countries. In some areas, births may be as high as 42 per 1000 population, and deaths around 12 per 1000. The difference, 30 per 1000, equals a 3% annual population growth. The resultant populations are comparatively young, with a median age in the upper teens.

*Infant and child mortality.* There are many ways to group children by age, including toddler, preschooler, school-age, adolescent, and so on. The greatest burden of illnesses and deaths falls on young children, commonly categorized into infants (ages 0 to 1 year) and under-fives (aged 0 to 4 years). Infants are subdivided further into neonatal (0 to 28 days) or postneonatal (29 to 365 days).

The best reported infant mortality rate (IMR), in Japan and parts of Scandinavia, is on the order of 6 to 7 deaths per thousand live births. In the United States the IMR remains around 10, but this national figure includes rates as high as 26 in parts of the District of Columbia. According to World Bank estimates for 1990, the overall IMR in low-income countries other than China and India was about 92 deaths per thousand live births. In some countries the figure was far higher; for example: Nepal, 121; Sierra Leone, 147; Bangladesh, 105; Malawi, 149; Mali, 166 (World Bank 1992). In limited geographic regions or particular demographic or ethnic groups the IMR may be even greater.

Assuming that the figures are correct, they will apply only to carefully defined populations, even within limited geographic areas. As an example, Lima Guimarães and Fischmann (1985) reported an IMR of 12.6 in the most favored quartile of the "nonshantytown" population, and 163.4 in the poorest quartile of shantytown dwellers, both within a single city (Pôrto Alegre) in southern Brazil. It is crucially important to recognize the wide disparities among socioeconomic groups that may share the same streets in developing countries.

The cultural significance of infant mortality has been discussed by various authors, most of whom acknowledge that the death of an infant is considered a tragedy everywhere. However,

> There are many societies in which these tragedies have a ranking different from ours. There are groups which consider that the illness, death, or disability of an old person or a breadwinner, or even the death of a buffalo, may threaten the structure or survival of a family. There are others which may consider the presence of an extra mouth to be a mixed blessing. These realities cannot be viewed as right or wrong (Newell and Nabarro 1989).

Child survival programs have powerful advocates. Among the advantages of such programs are great emotional appeal and the ability to demonstrate a rapid reduction in infant mortality, brought about directly by relatively straightforward and inexpensive interventions. Dramatic reductions in infant mortality rates are more readily attained when the initial situation is really dreadful. It is easier to reduce the IMR from 150 to 50 than it is to go from 15 to 10.

Despite their demonstrated success in the short run, child survival programs are not without their critics. Many people have pointed out the futility of merely saving lives by technological means without attempting to overcome the underlying causes of infant deaths, and without attention to the welfare of those whose lives are extended. Newell and Nabarro (1989) speak of "resources which are presently being directed (or misdirected) towards time limited and vertical infant anti-death programmes." The demographic and environmental implications of child-survival programs have stimulated commentators such as King (1990a, 1990b) to advocate that they not be undertaken without compensatory sustaining measures such as family planning. The ethical basis for these views is discussed further in chapter 3.

*Infectious diseases and immunization.* Children are the primary targets for immunization in all countries, and for good reason. Approximately 14 or 15 million children under five die annually (Gwatkin 1980), roughly 40,000 each day. Perhaps 60% of these deaths are associated with malnutrition; 80% are associated with diarrhea, with acute respiratory infections including pneumonia (often due to *Streptococcus pneumoniae, Haemophilus influenzae,* or pertussis), or with measles, tetanus, and malaria.

One major health problem in many poor areas is neonatal tetanus, which has sometimes been called "the forgotten disease." Its victims die so early that they never appear in vital statistics registers and never have the chance to be included in surveys of diarrhea, measles, or other childhood diseases.

Many children are permanently affected by infectious diseases, including an unknown number with neurologic damage from measles and other viral infections, and others with lifelong respiratory deficits. "The development of valvular heart disease as a sequel to rheumatic fever provides a clear example of how an infection in early life may have serious long-term effects" (Martyn 1991). The greatest social benefit from childhood immunization may well come from the protection of persons in their economically productive years. Vaccines that prevent such diseases as hepatic cancer in adults are important for the well-being of the entire community.

The distribution of causes of death in under-fives in three countries is shown in Figure 2–1. Note that the diseases preventable by current vaccines account for a minority of deaths in all three.

Greenwood et al. (1987) made an elaborate effort to identify, as precisely as possible, the causes of death in a series of 184 children who died in an area of The Gambia in 1981. Their findings, shown in Tables 2–3 and 2–4, demonstrate the significance of environmental factors such as season. The number of deaths from the six major immunizable diseases (diphtheria, pertussis, tetanus, measles, polio, and tuberculosis) is small, primarily because of a successful immunization program in this area. Although immunization coverage was better than that in the

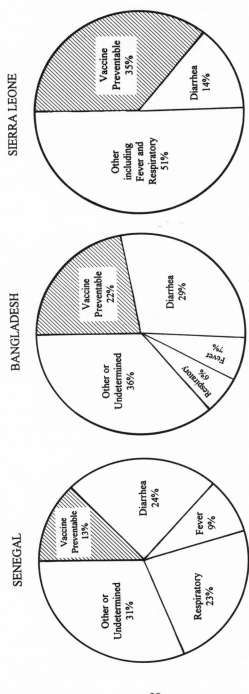

FIGURE 2–1. Cause of death in children 0 to 4 years old, in Senegal, Bangladesh, and Sierra Leone, in recent years. From Bart and Lin 1990.

TABLE 2–3.   Presumptive Causes of Death in 184 Children Under
7, The Gambia, 1981

| Diagnosis | Number | Percent |
|---|---|---|
| Neonatal causes | 42 | 23 |
| Acute respiratory diseases | 27 | 15 |
| Malaria | 25 | 14 |
| Meningitis and meningococcemia | 17 | 9 |
| Malnutrition and/or chronic diarrhea | 16 | 9 |
| Acute gastroenteritis | 14 | 8 |
| Measles | 6 | 3 |
| Septicemia | 6 | 3 |
| Chronic respiratory diseases | 5 | 3 |
| Miscellaneous | 7 | 4 |
| Unknown | 19 | 10 |
| Total | 184 | 100 |

Source: Greenwood et al. 1987.

TABLE 2–4.   Relationship Between Season and Cause of Death in 184 Children
Under 7, The Gambia, 1981

| Diagnosis | Number of Deaths | | Ratio, Rainy: Dry |
|---|---|---|---|
| | Rainy Season | Dry Season | |
| Malaria | 24 | 1 | 24.0:1 |
| Malnutrition and/or chronic diarrhea | 14 | 2 | 7.0:1 |
| Acute gastroenteritis | 12 | 2 | 6.0:1 |
| Neonatal | 32 | 10 | 3.2:1 |
| Acute respiratory | 14 | 13 | 1.1:1 |
| Meningitis | 3 | 14 | 0.2:1 |
| Others and unknown | 31 | 12 | 2.6:1 |
| Total | 130 | 54 | 2.4:1 |

Source: Greenwood et al. 1987.

Rainy season = June to November; dry season = December to May

United Kingdom, the overall infant mortality rate was 124 per thousand live births, five times greater than in the U.K. Malaria, diarrhea, and acute respiratory infections accounted for most of the disproportion.

The eradication of smallpox in the late 1970s showed the indisputable value of vaccines in reducing illness. More recently, the number of children disabled by poliomyelitis has declined so sharply that some authorities are predicting eradication of polio before the year 2000. Although cases continue to occur in other WHO regions, the Region of the Americans (the Western Hemisphere) reported only nine cases of flaccid paralysis due to polio in 1991, the last of which was in Junín, Peru on August 23, 1991. If three years pass without another confirmed case, the region can be certified free of wild poliovirus by a commission established for that purpose by the Pan American Health Organization.

*New vaccines for the Expanded Programme on Immunization (EPI).* Table 2–5 shows the currently recommended series of EPI immunizations. The EPI schedule, established in 1975, was never intended to be immutable, and over the years changes have been made to suit local conditions. In certain endemic areas, other vaccines such as Japanese encephalitis are added. According to the EPI (1992a),

> Yellow fever vaccine should be routinely administered to children under one year of age in all countries at risk for yellow fever by 1993. Hepatitis B vaccine should be integrated into national immunization programmes in all countries with a hepatitis B carrier prevalence (HBsAg) of 8% or greater by 1995 and in all countries by 1997. Target groups and strategies may vary with the local epidemiology. When carrier prevalence is 2% or greater, the most effective strategy is incorporation into the routine infant immunization schedules. Countries with lower prevalence may consider immunization of all adolescents as an addition or alternative to infant immunization.

Maynard et al. (1989) calculated that deaths from the long-term consequences of hepatitis B infection are equal to those from acute cases of measles, 50% greater than pertussis, three times greater than tetanus, and six times those from polio. Hepatitis B infection "differs from other EPI-targeted diseases in that the majority of the related morbidity and mortality occurs during the adult period, following perinatal or childhood infection that is largely asymptomatic" (Maynard et al. 1989). Field trials such as that of Schoub et al. (1991) in South Africa have been conducted to integrate this vaccine into the normal EPI series. The region possessed what the authors term as "excellent" primary health care facilities, and 93% of children received three doses of polio and DPT vaccines. Aftr three months of meetings, conferences, pamphlets, and radio messages, "acceptance by both health care workers and the population seemed to be excellent." Nevertheless, immunization coverage for the hepatitis B vaccine dropped from 99% for the first dose to 39% for the third, and fewer than 7% of children were immunized on schedule.

> The high dropout rate in an otherwise excellent primary health care programme emphasizes the difficulties in introducing a new vaccine to routine (EPI) programmes and the need to devise strategies to minimize disruption. . . . A tetravalent vaccine for diphtheria, tetanus, pertussis, and hepatitis B, with all four constituents mixed into the same phial, would ensure that hepatitis B vaccine coverage would at least reach that of diphtheria, tetanus, and pertussis (Schoub et al. 1991).

TABLE 2–5.   Recommended Schedules for Immunization of Normal Infants and Children, United States, and for EPI Vaccines in Developing Countries, 1992–1993[a]

| Age | United States | Developing Countries |
|---|---|---|
| Birth | HBV | BCG, OPVO, HBa1[n] |
| 6 weeks | | DPT1, OPV1, HBa2, HBb1 |
| 1–2 months | HBV | |
| 2 months | DTP1, HbCV1, OPV1 | |
| 10 weeks | | DPT2, OPV2, HBb2 |
| 14 weeks | | DPT3, OPV3 |
| 4 months | DTP2, HbCV2, OPV2 | |
| 6 months | DTP3 (HbCV3[h]) | measles in some areas[e] |
| 6–18 months | HBV | |
| 9 months | measles[d] | measles, YF[Y] |
| 12 months | (HbCV3[h]) | |
| 15 months | MMR, (HbCV4[h]) | |
| 15–18 months | DTaP or DTP, OPV | |
| 4–6 years | DTP or DTaP, OPV | |
| 11–12 years | MMR | |
| 14–16 years | Td | |
| Adult women | | TT (childbearing age) |

Source: Peter, Georges: Childhood immunizations. NEJM 327 (25) Dec. 17, Pp. 1794–1800, 1992. Reprinted by permission of the New England Journal of Medicine and EPI (unpublished).

Abbreviations: BCG, Bacille Calmette-Guérin (for Tuberculosis); DPT, Diphtheria toxoid, Pertussis whole cell vaccine, Tetanus toxoid (WHO terminology); DTP, the same (U.S. terminology); DTaP, Diphtheria toxoid, Tetanus toxoid, Acellular pertussis vaccine; HbCV, *Haemophilus influenzae* type b Conjugate Vaccine; OPV, Oral Polio Vaccine; MMR, Measles, Mumps, and Rubella; Td, Tetanus toxoid, full dose, and Diphtheria toxoid, reduced dose, for adults; TT, Tetanus toxoid.

[a]Consult the originals for cautionary notes and further details. Consult recent issues of the MMWR for current U.S. recommentations and the Weekly Epidemiological Record for WHO recommendations.

[d]In counties at high risk of measles. Use single-antigen measles vaccine for children aged under 1 year, and MMR for children 1 year or older.

[e]In most areas.

[h]Depending on manufacturer and product.

[n]Schedule a, in areas where perinatal transmission of hepatitis B is common; schedule b in places where early transmission is not a problem.

[y]Yellow fever vaccine should be routinely included in countries at risk in Africa.

Some consequences of the inclusion of hepatitis B vaccine in the EPI series are shown in Table 5–13.

Table 2–6 lists additional vaccines predicted by a committee of the Institute of Medicine, U.S. National Academy of Sciences, to become available by 1996, but many of the listed vaccines are unlikely to meet that target. Note the absence of an AIDS vaccine, which was not on the screen in those days.

TABLE 2–6.    Vaccines for International Use Predicted by the Institute of Medicine in 1986 to be Available within a Decade

| Pathogen | Vaccine | Years to Licensure |
|---|---|---|
| Dengue virus | Attenuated live vector virus | 10 |
| *Escherichia coli* (enterotoxigenic) | Purified antigens | 10 |
| | Attenuated, engineered | 10 |
| Japanese encephalitis virus | Cell-culture grown, inactivated | 6–8 |
| *Mycobacterium leprae* | Armadillo-derived | 8–10 |
| *Neisseria meningitidis* | Conjugated polysaccharides, groups A, C, Y, W135 | 4–6 |
| *Plasmodium* species | *P. falciparum* synthetic or rDNA sporozoite antigen | 5–8 |
| | Multivalent synthetic or rDNA sporozoite antigens; *P. falciparum, P. vivax, P. ovale, P. malariae* | 8–10 |
| Rabies virus | Vero cell-grown, inactivated | 3 |
| | rDNA glycoprotein | 3 |
| | Live vector virus with glycoprotein gene | 3 |
| *Salmonella typhi* | Ty21a mutant | 1 |
| | Auxotrophic mutant | 5–8 |
| *Shigella* species | Plasmid-mediated determinants | 10 |
| *Streptococcus* A | Synthetic M protein | 6–8 |
| *Streptococcus pneumoniae* | Conjugated polysaccharides | 5 |
| *Vibrio cholerae* | Genetically defined live mutant | 5–7 |
| | Inactivated antigens | 3–5 |
| Yellow fever virus | Cell culture grown, attenuated | 2–4 |

*Source:* Jordan 1989.

Note: Excludes *Haemophilus influenzae,* hepatitis A and B, parainfluenza, respiratory syncytial viruses, and rotaviruses listed on other tables in the original publication. Consult original for details.

## Population

*Population distribution.* Population density varies greatly, from about 1 person per square kilometer in Mauritania to almost 4,500 in Singapore. The number and proportion of people living in urban areas are increasing rapidly, particularly in the encircling slums that acocunt for up to two-thirds of the total population of some tropical cities.

In 1950 about 26% of all people lived in urban areas, in 1970 about 35%, and at present more than half of the world's population lives in cities of 100,000 or more. The urban concentration of people is significant epidemiologically, considering the spread of infectious agents, including sexually transmitted pathogens, fecal and industrial pollution, alcoholism, crime, and stress. On the other hand, the efficiency of certain public health measures is enhanced by higher population density.

It is likely that overall mortality is lower in the cities than in the countryside of developing countries. The widespread opportunity to earn regular wages and resultant improvements in individual and community nutrition, education, housing, environmental sanitation, and medical care can reduce the incidence of nutritional, infectious, and vector-borne disease in industrialized urban dwellers compared to residents of rural villages.

*Population increase and environmental decline.* Most people consider economic development a good thing because it offers the opportunity of a better life to the majority of the population. Nevertheless there are many critics, including an increasing number who are concerned with the corresponding impoverishment of the environment. Rapid population increase is thought by many to be the chief culprit in this process, and therefore the greatest menace facing humanity (see also chapter 4). Increases in agricultural production necessitated by population growth will lead to still higher total levels of environmental pollution from fertilizers and pesticides. The application of biotechnology to these problems, by development of higher-yielding and pesticide-resistant food crops, may eventually have greater influence on human health than those innovations, such as vaccines, directed specifically at disease control.

## Primary Health Care

In 1977 the 30th World Health Assembly of the World Health Organization, spurred by representatives of the developing countries, resolved that the primary target of the WHO and of its member governments by the year 2000 should be "the attainment by all citizens of the world of a level of health that will permit them to lead a socially and economically productive life. This has become known as "Health for All by the Year 2000," or simply HFA2000.

Organized by WHO and UNICEF, the countries of the world convened a major international conference on Primary Health Care (PHC) in September, 1978. Held at Alma Ata, USSR, the conference was attended by representatives of 134 governments and numerous nongovernmental organizations. It was the largest international gathering on health ever convened. In the resulting *Declaration of Alma Ata* the signatory countries defined PHC as:

> essential health care based on practical, scientifically sound and socially acceptable methods and technology made universally accessible to individuals and families in the community through their full participation and at a cost that the community and country can afford to maintain at every stage of their development in a spirit of self-reliance and self-determination. It forms an integral part both of the country's health system, of which it is the central function and focus, and of the overall social and economic development of the community.

The basic elements of Primary Health Care are:

- Immunization against the major infectious diseases
- Maternal and child health care

- Efforts for the prevention and control of locally endemic diseases
- The provision of essential drugs
- Education concerning prevailing health problems and the methods of preventing and controlling them
- The promotion of food supply and proper nutrition
- An adequate supply of safe water and sanitation

## Major Programs Associated with Primary Health Care

In the period since Alma Ata most developing countries have adopted at least some elements of PHC. Some of the large-scale high-impact primary health care intervention programs include:

- The "Child Survival and Development Revolution," sponsored by the United Nations Development Programme (UNDP), UNICEF, the World Bank, WHO, and the Rockefeller Foundation through their Task Force for Child Survival established in 1984
- The U.S. Agency for International Development's CCCD (Combating Childhood Communicable Diseases) program in a number of poor African countries
- The "Safe Motherhood Initiative" of the World Bank and others
- UNICEF's emphasis on GOBI (growth monitoring, oral rehydration therapy, breast-feeding, and immunization), sometimes combined with FFF (food supplementation, female literacy, and family planning)

*Goals and targets.* In 1988 the Task Force for Child Survival sponsored a conference in Talloires, France, which announced a set of goals particularly relevant to the capabilities of biotechnology. The resultant *Declaration of Talloires* proposed to accomplish the following, by the year 2000:

- Poliomyelitis: global eradication
- Measles: 90% reduction in cases and 95% reduction in mortality compared to preimmunization levels
- Neonatal tetanus: virtual elimination
- Diarrhea: 25% reduction in incidence and 70% reduction in the 1980 under-five mortality level
- Acute respiratory infections: 25% reduction in current under-five deaths
- Infant mortality: reduction by half or to 50 per thousand live births
- Under-five mortality: reduction by half or to 70 per thousand children
- Maternal mortality: 50% reduction

*Criticisms of global programs.* Programs such as the EPI, Control of Diarrheal Disease (CDD), the Child Survival Revolution, and Safe Motherhood have generally been greeted with near-unanimous support by delegations attending large international meetings. However, some health authorities comment privately that while it is hardly possible to vote openly against such humanitarian resolutions, they and their governments retain a skeptical attitude. The first decade of HFA2000 was typified by genteel and temperate discussion, but around the midpoint between Alma Ata (1978) and the year 2000, reservations were expressed with greater frequency. The Minister of Health of a large African country, speaking at a conference on international health in 1989, criticized GOBI "with or without the FFF" as an unwelcome diversion from the thrust of national health care policies. A senior educator from a large South Asian nation characterized Child Survival and GOBI as poorly conceived mass campaigns and simplistic approaches to complex problems. He claimed that poor people were "the target of magic bullets fired by foreign bureaucrats, who have missed the target." His criticism of "external consultants who teach the heathen the way to heaven" are not new. The point is that the foreign biotechnologist should be aware that his or her innovation may not be universally welcomed, for reasons quite extraneous to its intrinsic merits.

*Alternative conceptions of PHC: horizontal and vertical.* From the beginning of large scale official development assistance (ODA), commonly referred to as "foreign aid," donors have supported monovalent or "vertical" programs that target (1) *diseases* such as malaria, trachoma, yaws, or leprosy; (2) *problems* such as malnutrition or childhood diarrheal diseases; and (3) *demographic groups and substantive areas* such as maternal and child health.

Soon after the Alma Ata conference, two factions became identifiable. One favored "selective PHC," a strategy intended to increase the effectiveness of PHC expenditures. The selective approach establishes health priorities in relation to their prevalence, morbidity, mortality, cost, and negative impact on the community. Both needs and results should be readily quantifiable. Where control methods exist and are accessible, specific diseases or identified problem areas are attacked by the most cost-effective means available.

A contrasting viewpoint is held by proponents of "comprehensive PHC." They believe that health is indivisible, community participation is essential in planning and implementation, and control should rest with the people affected and not with outside experts. Individuals who adhere to this comprehensive, horizontal approach view the Alma Ata declaration as empowering local people to make decisions about their own needs and the means to achieve them.

The selective and comprehensive approaches are fundamentally opposite views, which affect the readiness with which people are willing to accept new biomedical (and other) procedures and the appropriateness of particular communities as field sites for clinical trials. Associated concepts of equity, social justice, and right to health constitute powerful nuclei of concern and activism for many individuals and organizations.

## Resource Allocation to PHC

The Alma Ata declaration subsumed for PHC the lion's share of attention as "an integral part both of the country's health system, of which it is the central function and focus, and of the overall social and economic development of the community."

The global high-impact programs listed earlier have common objectives, a demand for long-term commitment, and an appetite for resources. Where funds, trained personnel, and physical facilities are minimal, the adoption of such goals and the programs that accompany them may place extreme stresses on the local capacities.

While the *Declaration of Alma Ata* gives responsibility for implementation to governments, the Ministry of Health in very poor countries may find this difficult to do. In such cases most of the actual work on the ground has been carried out by external donors or by private voluntary organizations either acting independently or as contractors for development agencies. In these situations the biomedical innovator must deal not only with national, regional, district, and local governmental officials, but with a web of third or fourth parties in arranging for field trials of new products. Such arrangements are discussed further in chapter 4.

## Paying for Primary Health Care

"There is a myth that primary health care is a cheap solution to a difficult and complex problem. Not so, and those who attempt to sell it as a cheap solution do it a poor service" (Fendall 1985). Even in the best of times, locally generated funds are inadequate to pay for health and other social services in most parts of the world. In the poorest countries the burdens added by the debt crisis, described earlier, have added to the need for subsidization by foreign donors. Many public health services, particularly vertical programs financed by outside donor agencies, are short-lived and terminate with few durable results. Programs based on mobilization of indigenous sources may have greater hope for longevity, if not for permanence. A cottage industry of consultants, analysts, advisors, and advocates has emerged around the many different schemes proposed for community financing of primary health care.

National-level funding sources include general tax revenues or employer/employee-financed social security programs. Some funds may be raised through prepayment of compulsory or voluntary insurance premiums. Fee-for-service schemes also exist, in which full cost recovery is often impossible, and even partial recovery of system costs is difficult. Affordable charges may be limited to the range of five to twenty U.S. cents. Administrative costs to collect, record, safeguard, and transfer such small fee-for-service payments may equal or exceed the revenues generated. In such settings the cost-effectiveness of biomedical innovations is put to the acid test.

*The Bamako Initiative.* In 1987 a meeting of the WHO Regional Committee for Africa was held in Bamako, Mali. The ministers adopted a plan, known formally as "The Bamako Initiative: Women's and Children's Health through the Funding and Management of Essential Drugs at the Community Level." The aim was to help assure availability of drugs in primary health care facilities and the continuity of PHC and maternal and child health programs. The essence of this plan is to provide "seed drugs" to health institutions, from stocks purchased in bulk by national ministries or external donors. These are sold to patients at a small mark-up, which pays for the drugs, establishes a revolving fund for future purchases, and raises a small amount of money to support maternal and child health

(MCH) services at the health center. The scheme, based on pilot projects in Benin and Guinea, has generated much debate.

Supporters of the Bamako Initiative point out that most Africans have a tradition of paying for health care, whether from local or western practitioners; that several countries routinely charge a fee for services; and that in any case, many people must obtain their drugs from private pharmacies. Some reports (e.g., Salako 1991 in Nigeria) describe effective operation of such a scheme.

On the other hand, the plan has been criticized on grounds of equity. Many people are unable to pay even minimal charges, particularly in the preharvest season when there is little available cash, nutritional status is low, and infectious diseases may be most frequent. Therefore some system of insurance has been suggested, in which a family may prepay after the harvest to cover the cost of services provided later in the year. Many obvious difficulties are associated with such a plan.

## Primary Health Care and Political Risk

Primary health care programs are not devoid of political risk. Some governments in developing countries may find community organization for comprehensive PHC menacing, particularly in areas with current or threatened insurrections. Community health workers have been killed because of their dedication to community empowerment and the precepts of primary health care. On the other hand, governments may utilize actual or feigned PHC workers as agents for infiltration and surveillance, or as vehicles for social control. The foreign innovator should understand these issues before considering a field trial in a politically sensitive area.

## Summary of Problems of Health Services in Developing Countries

Specifics vary, but health and medical services in most poor countries tend to include many of the following features:

- The government is well-intentioned toward social justice and the welfare of its citizens, but gives little priority to health ministries in the form of financial, material, and human resources.
- Confidence by the people in the government in general, and in health services in particular, is low.
- Policies for personnel recruitment, training, compensation, and support do not induce the needed level of staff efficiency, morale, commitment, or retention.
- Programs are curative in orientation and do not attack the root causes of ill health.
- Consumable items are in short supply, and preventive maintenance and repair of equipment and facilities is inadequate or lacking.
- Communities are not consulted about their concerns or helped to solve their own problems.
- Resources of all kinds are concentrated in one or a few urban areas.
- The poorest, most distant, most vulnerable, and neediest group are the least served.
- Education, particularly of women, is inadequate or misdirected.
- There is excessive dependence on, and influence by, foreign donors.

## The Cultural Context

The biotechnologist who wants his or her product used in developing countries must acknowledge that others may not give equal significance to "health," nor to the diseases considered important in the laboratory back home.

> Many health personnel have viewed the issues in immunization as just providing the goods: produce effective vaccines, distribute them, maintain the cold chain, announce when and where vaccinations are being provided, and administer the vaccine—the assumption being that, if immunization is supplied, people will of course recognize it as an absolute good and desire it (Pillsbury 1990).

### *Conceptions of Illness*

Biomedical laboratory scientists may feel uncomfortable at having to deal with anthropological approaches to health. Like the fish unaware of water, we may not reflect upon the controlling influence of our own culture, in which individual initiative and achievement are rewarded, and science and technology are used to attain mastery over nature. In contrast, the daily lives of many people throughout the world are directed by ideas of harmony with nature rather than mastery over it, and with the constant need to carry out (or abstain from) certain behaviors in order to avoid the penalties of illness or ill fortune that are certain to befall transgressors.

Cultures differ in their ideas about the varieties, categories, and nature of illnesses, how and why they occur, and what can be done to prevent or cure them. The catalog of diseases familiar to western scientists reflects a certain world view not necessarily shared by all people. Even in our scientific society, so-called experts disagree about the relative roles of heredity and environment in causation of diseases. They question the significance of tobacco, dietary cholesterol, sedentary behavior, and other risk factors (analogous to sins and taboos) in causation of cancers, atherosclerosis, and other chronic conditions.

In examining the antecedents of an illness, we usually ask, How did it happen? In traditional societies people ask, Why did it happen? In a comprehensive review, Murdock (1980) found that concepts of causation are either natural (i.e., appear reasonable to medical science: infection, stress, deterioration, trauma), or supernatural. The latter may (a) be mystical, with an impersonal cause such as fate or bad luck; (b) result from ominous sensations such as dreams; (c) emanate from contact with a contagious object, substance, or person; or (d) indicate retribution for some moral violation. Other supernatural causes of illness include the behavior of a soul, ghost, spirit, or god, or the acts of an envious or malicious person, by sorcery or witchcraft. In addition, the idea of universal opposites such as yin and yang and the need for a balance of "heating" and "cooling" foods, herbs, medicines, activities, and illnesses, is also considered important in many parts of the world. These attitudes are not trivial or merely picturesque. They are as meaningful in the management of daily life as are our own cultural beliefs and practices.

An example of differing perceptions is found in the field of immunization.

Whereas the purpose, and the benefits, of immunizations should be self-evident to educated Western people, others may not share that viewpoint. Vaccinations are difficult to sell, especially to healthy people who may be unwilling to spend the money or time, or to risk adverse consequences, in order to be protected against a disease that they or their children may never get.

*Misperceptions about immunization held by village people.* Pillsbury (1990) has assembled a valuable collection of widely held misperceptions, which is of interest to anyone considering an immunization program among people in developing countries.

1. Misperceptions Related to Prevention

   - *Immunization is curative.* In Honduras many parents believe that immunization is useful only for curing diseases. Some do not understand the concept of prevention.
   - *Healthy children do not need immunization.* In Indonesia and Nigeria parents see no need to bring healthy children for immunization. Some (reasonably) do not want to expose their healthy children to possible complications of the immunization.

2. Misperceptions about Contraindications

   - *A sick child should not be immunized.* This is a widespread idea among mothers and health workers in many different countries. Some parents in India believe that a weak child cannot stand the shock of a powerful health injection.
   - *The belief in climatic or seasonal contraindications.* In India and Sri Lanka rainy days are considered a bad time to receive an immunization; in Sri Lanka the summer is a bad season for immunization because of the abundance of sour, "heat-producing" fruits that unbalance the blood.
   - *Immunization will lead local gods to inflict misfortune on the household.* In India and Nigeria measles is believed to be supernaturally caused, and immunization may anger the disease-causing god or goddess.
   - *Women should not be immunized while pregnant.* This common belief reduces the acceptance of tetanus toxoid by pregnant women.

3. The Misperception that Immunization is Ineffective

   - *Beliefs due to ineffective vaccines.* "Whenever children have been immunized (or are believed to have been immunized) against a disease and then later contract it anyway, community confidence in immunization is shaken."
   - *Beliefs due to false expectations.* Parents may falsely believe that immunization protects a child against diarrhea, dysentery, vomiting, fever, pneumonia, malaria, and even coughs and colds, and are disappointed when those conditions occur. In Haiti mothers may think that immunization strengthens resistance to illness in general.
   - *Immunization may be confused with other injections.* Mothers may confuse oral polio vaccine with vitamin drops given in a similar way, and they may not distinguish between vaccines, antibiotics, or tonics given by injection.

4. Other Misperceptions

   - *Vaccinations are given to small children to make them become accustomed to taking western medicines.*

- *Vaccination of children reduces their fertility later in life.*
- *Tetanus toxoid given to pregnant women is covertly related to family planning; it will make the babies large and deliveries difficult.*

*Misperceptions held by Western scientists.* Misperceptions about immunization do not occur just among uneducated village people. The Advisory Committee on Immunization Practices (1989) of the U.S. Public Health Service has prepared a list of erroneous contraindications accepted by health practitioners in the United States (Table 2–7).

Health professionals in other countries also have inappropriate reasons for not immunizing children. In some cases these missed opportunities have unfortunate outcomes, as in an outbreak of paralytic poliomyelitis among unimmunized infants in Bulgaria in which "Temporary contraindications recognized by physicians . . . (fever, upper respiratory infections, etc.) were an important reason for delayed immunizations" (Weekly Epidemiological Record 1992).

Western scientists also exhibit misperceptions about the people who may be the recipients of their innovative technologies. Almost 30 years ago Polgar (1962; paraphrased by Heggenhougen and Clements 1987) described four common fallacies of approach by Western health professionals working in developing countries. These have not changed appreciably in the intervening decades:

1. *The fallacy of the empty vessel.* Communities are often approached as if they are empty vessels with no prior concepts of health and illness and means of alleviating ailments prior to the introduction of allopathic medicine. This fallacy is well recognized, but nevertheless still present in program planning and implementation.

TABLE 2–7. Commonly Accepted Erroneous Contraindications against Vaccination

---

- A reaction to a previous dose of DTP vaccine with only soreness, redness, or swelling in the immediate vicinity of the vaccination site or a temperature of <40.5C (105F).
- Mild acute illness with low-grade fever or mild diarrheal illness in an otherwise well child.
- Current antimicrobial therapy or the convalescent phase of illness.
- Prematurity (the appropriate age for initiating immunizations in premature infants is the usual chronological age; vaccine doses should not be reduced for preterm infants).
- Pregnancy in the mother or another household contact.
- Recent exposure to an infectious disease.
- Breast-feeding (the only vaccine virus that has been isolated from breast milk is rubella vaccine virus; no evidence indicated that the breast milk of women immunized against rubella is harmful to infants).
- A history of nonspecific allergies or relatives with allergies.
- Allergy to penicillin or any other antibiotic agent, except anaphylactic reactions to neomycin (e.g., MMR-containing vaccines) or streptomycin (e.g., OPV). (None of the vaccines licensed in the United States contain penicillin).
- Allergies to duck meat or feathers (no vaccine available in the United States is produced in substrates containing duck antigens).
- A family history of convulsions in children who require vaccination against pertussis or measles.
- A family history of sudden infant death syndrome in children who require DTP vaccination.
- A family history of an adverse event, unrelated to immunosuppression, after vaccination.

---

*Source:* Immunization Practices Advisory Committee 1989.

2. *The fallacy of the single pyramid.* Some wrongly believe that a single hierarchical structure exists within all communities and that success will be achieved merely by providing the "necessary" information to the perceived head or chief. In reality, communities are complex entities composed of many grids of formal and informal relationships and obligations.

3. *The fallacy of interchangeable faces.* It is naively assumed that one village is much like every other village—and that if a particular approach works in one place it will work elsewhere.

4. *The fallacy of the separate capsule.* It is believed that health status is not affected by anything but health care interventions. Even though intersectoral approaches to health are always discussed, they are not often practiced.

*Inappropriate failure to immunize by health workers in developing countries.* Many health workers in clinics are afraid to immunize anyone but a "healthy child" (Henderson 1984). Many children are brought to a hospital only when severely malnourished or otherwise ill. These visits present an opportunity to immunize because the child is not likely to be brought back to the clinic after recovery. Health workers should be informed that the chance of adverse reaction to vaccines is low while potential benefit is high. Ofosu-Amaah (1983) writes that in their concern about contraindications, refusal by overzealous health workers to vaccinate mildly ill or malnourished children decreases acceptance of the immunization program.

Foreigners sometimes view villagers in developing countries as stubbornly resistant to trying new things. For people living on the margin, for whom survival is a daily challenge that may be jeopardized by a mistake in judgment, and for whom suspicion of outsiders may be soundly based, skepticism about the possible hazards of unknown procedures promoted by strangers with unknown motivations may be a highly justifiable attitude.

A fair predictor of attitudes toward biomedical innovations is a group's acceptance of current public health practices. Community-wide programs such as vector control are rarely subject to individual discretion. In the case of diagnostic procedures, persons in the community similarly have little knowledge or control over selection of methods. Where individuals do have a choice, some, for whatever reason, may elect not to participate in programs that bring piped water, sanitary facilities, educational sessions, or messages through mass media. Much attention has been given to public health procedures that require direct individual initiative, such as the use of oral rehydration salts for diarrhea, or participation in immunization programs. Questions about incentive and motivation must be addressed to understand how and why persons in various communities relate to such activities.

## Individual Acceptance of or Resistance to Innovation

Receptivity to innovations is rooted in culture and related to attitudes toward taking risks in general. Crouch and Wilson (1982) have listed some factors, paraphrased here, that can influence judgments of risk:

1. *Immediacy of effect.* Are the consequences immediate or delayed? If there is a latent period, how does it affect perception of the innovation?

2. *Availability of alternatives.* Are there any alternatives that will produce similar benefits with reduced risks? Is there a choice?
3. *Knowledge about risk.* Are the people properly informed about the benefits, risks, and real purposes of the innovation? Is there some kind of plot or conspiracy behind it? If it is legitimate, how large are the uncertainties?
4. *Necessity of exposure.* Is exposure to the innovation necessary, coerced, or voluntary?
5. *Familiarity of risk.* Old and familiar risks are accepted more readily than new and strange ones.
6. *Distribution of risk.* Will everyone be exposed equally to the innovation or are only some selected (e.g., for trials or experiments)?

Other factors that affect acceptance of immunizations include:

- Parental age and education
- Mobility and isolation of populations
- Lack of information on the part of parents (a function both of the parents themselves, and of the information disseminated)
- Degree of faith in "science"
- Previous frustrating experiences with immunization
- Support (or lack of support) by influential members of the community
- Involvement of mother in local social and self-help organizations
- Family tendency to take children to clinics
- Favorable concept of injections as preventive and curative
- Familiarity with the disease and knowledge of its causality
- Perceived susceptibility to the disease by oneself or one's child
- The severity or consequences of having the disease
- Perceived hazards, drawbacks, and disadvantages of immunization

The comprehensive list of reasons given by mothers for not immunizing their children (Table 2–8) should be required reading for every biomedical scientist interested in making a vaccine for use in developing countries.

Despite these many reasons for so-called "noncompliance," it is possible to obtain a high degree of acceptance even in a community with a dubious view of the immunization.

In sum, individual participation is determined by a large number of factors, some of which can be adequately anticipated. Other factors determining individual acceptability are specific to a given village or region. No template exists for an ideal and universal approach.

## Organized Activities For or Against Immunization

*Inherent community resitance.* Imperato (1975) described poignantly how "jet injectors" were viewed with terror by villagers in the Bambara-speaking region of Mali, where residents made elaborate plans to avoid immunization, sometimes rehearsing for weeks to escape from the public health teams.

It is not only in developing countries that certain communities and subgroups are antagonistic to immunization, or to other public health practices. Everywhere, members of some religious sects object to immunization on theological grounds

TABLE 2–8.   Reasons Given by Mothers for Not Immunizing Their Children

A. Reasons related to characteristics of the mother and other caretakers:
  1. Time constraints and competing priorities
    a. Meeting subsistence needs is more essential
    b. Other economic activities have higher priority
    c. Family problems consume large amounts of time
  2. Other socioeconomic constraints
    a. Older children restrict mobility but cannot be left behind alone
    b. Lack of clean or proper clothes
  3. Lack of knowledge about immunization
    a. Lack of knowledge about kinds of vaccines and how vaccination works
    b. Lack of knowledge about schedules and repeat doses
    c. Beliefs about contraindications for vaccination
  4. Low motivation for immunization
    a. Health is a relatively low value
    b. Unconvinced of importance of immunization
    c. Unconvinced of efficacy of immunization
    d. Discouragement over continuing poor health of child despite efforts to make the child thrive
    e. Fatalism about child survival
    f. Maternal "negligence" (too many children or the child is a girl)
  5. Fears
    a. Fears of side effects
    b. Fears of criticism or other unsupportive comments
    c. Fear that health workers will apply pressure to use family planning
    d. Fear that vaccination is a covert form of family planning or that it might result in sterilization
    e. Fear that acceptance creates indebtedness to the vaccinator
    f. Public shyness or embarrassment
  6. Community opinion
    a. Negative opinions within the community about immunization
    b. Esteemed community leaders have not advised or told community members to immunize their children
B. Reasons related to characteristics of the vaccines:
  1. Side effects of the vaccines
  2. Belief that vaccine is not effective
C. Reasons related to characteristics of the delivery of immunization services (including patient-provider interaction):
  1. Accessibility
    a. Immunization site is too far away
    b. Road or path to clinic is hazardous or impassable after rain or in the rainy season
  2. Availablity
    a. Immunization services are scheduled at a time that conflicts with other duties
    b. Not adequately informed of the immunization schedule
    c. Arrival of vaccinators is unpredictable (e.g., due to weather or distance)
    d. Vaccines not in supply (or staff won't open a new vial)
    e. Some vaccines systematically not used on certain days of the week
  3. Acceptability
    a. Poor treatment from health staff (experienced personally or reported by others in the community)
    b. Poor injection technique causing pain or side effects
    c. Mistrust of government vaccinators
    d. Curative services are not provided (nor material aid)
    e. Facility is overcrowded: long wait, service rushed       continued

TABLE 2–8.   (*continued*)

---

   4. Affordability
       a. Direct costs
       b. High opportunity cost of the mother's time
D. Reasons related to communication to the public about immunization:
       a. Inadequate communication about the nature and benefits of immunization
       b. Inadequate communication about when and where immunization is being provided
       c. Inadequate communication about when to return for follow-up doses.

---

*Source:* Pillsbury 1990.


TABLE 2–9.   Religious Groups Possibly Opposed to Immunization

---

Amish

Church of Christ in Christian Union

Church of Christ, Scientist

Church of the First Born

Church of God (several types)

Church of Human Life Sciences

Church of the Lord Jesus Christ of the Apostolic Faith

Church of Scientology

Disciples of Christ

Divine Science Federation International

Faith Assembly

Hare Krishna

Hutterites

Kripala Yaga Ashram

Mennonites

Netherlands Reform Church

Rosicrucian Fellowship

Worldwide Church of God

---

*Source:* Novotny et al. 1988.


and have become quite sophisticated about what is or is not tolerable. In New Zealand some parents, including Jehovah's Witnesses, will not accept blood-based vaccines, but will allow yeast-derived vaccines for their children (Milne et al. 1989). Table 2–9 lists some religious groups in the United States whose doctrines oppose immunization. Adherence to these religious tenets may have serious consequences. Between January and April, 1991, at least nine outbreaks of rubella, involving more than 400 cases, occurred in Amish communities in several states, among whom outbreaks of measles and pertussis have also been reported (MMWR 1991). In the Netherlands between September and November, 1992, at least 33 cases of paralytic poliomyelitis occurred among a sociographic cluster of people who refuse immunization for religious reasons.

## Biomedical Technology and Health

*Appropriate technology.* The subject of "appropriate technology" has been popular for several decades. Several organizations, such as Program for Appropriate Technology for Health (PATH) in Seattle, Washington, and Appropriate Health Resources and Technologies Action Group (AHRTAG) in London have long made major efforts in this direction.

Appropriate technology in the usual sense is technology that is relatively simple, labor-intensive, and does not require substantial capital investment.

Perhaps some terms should be borrowed from the computer world, such as "user friendly," to indicate procedures that are performed with relative ease and efficiency by operators having minimal knowledge and skill. The complexity of the underlying technology is irrelevant so long as the "interface" appears simple to the user. The most direct and cost-effective means to accomplish a specified goal in certain settings in developing countries could possibly employ recombinant DNA, monoclonal antibodies, or computer microchips, which would make them the most appropriate technology for the job.

*The historical record in the West.* The connection, or lack of it, between biomedical technology and health has given rise to a large body of literature. Some scholars have concluded that medical technology has had little effect on levels of health in the currently developed countries, and have credited the decline in general and childhood mortality to advances in agricultural practices, nutrition, water supply and sanitation, refrigeration, and housing. General improvements in social conditions, including enlightened labor and school legislation, and public health services in the broad sense, are seen widely as contributors to improved health and longevity. Commentators point to the marked reduction in deficiency diseases, scarlet fever, measles, diphtheria, and pertussis that began before the causative agents, or any means of prevention or treatment, were known. On the other hand, diseases such as rubella and polio declined sharply only after immunization became widely available.

*Attitudes toward biomedical technology.* Technology has always had its detractors. During the mid-1970s technology in general was severely criticized, and comments such as the following were commonplace:

> With regard to the developing countries, several authors have reviewed the available evidence and have been unable to find reason to believe that either preventive or curative medical efforts contributed significantly to the rapid decline in death rates since about 1930. . . . Illich (1976), Carlson (1976), and McKeown (1976) have put all this evidence together to present convincing arguments that medical technology is usually useless, frequently harmful, and a tremendous waste of public and private funds. . . . These views have had considerable influence. Most west-

ern donors have given a very low priority to health since the early 1960s, in large part because of a growing belief that preventive and curative medical efforts are irrelevant to improving the health and well-being of poor people in developing countries (McCord 1978).

Even in the United states there is considerable opposition to experimentation with recombinant DNA and the release of engineered organisms. The fear of unknown biomedical hazards is partly responsible, but objections have been raised for other reasons. For example, special reluctance has been expressed to the administration of genetically engineered bovine somatotropin, which stimulates cattle to produce up to 25% more milk. The primary argument is not physiological but economic, with small producers arguing that use of the hormone will worsen the existing surplus of milk and drive them from business (Schneider 1989a, b). Similar arguments are made for other commodities in developing countries. One commentator believes that biotechnology will cause economic havoc in areas that produce certain crops such as vanilla and sugar, currently obtained from labor-intensive natural sources. Cheap methods have been developed for artificial synthesis of these products and for cloned plants with much higher product yields.

> Biotechnology thus holds out a multiple threat: overproduction (of commodities that are not necessarily in short supply), substitution of natural by engineered products, the transfer of production from the fields of the third world to the laboratories and factories of the first, and the loss of bargaining power for producers of specific commodities. . . . Even more ominous, the biotech companies are now packaging their technology in a manner that favours vertical integration and monopolies. . . . The biotechnology revolution is thus likely to be considerably worse than the green revolution in its impact. . . . In the long run, biotechnology's potential for loading the dice even more heavily against the third world and making its terms of trade yet more adverse is frightful. . . . Clearly, the scenario is dismal. What makes it particularly worrisome is that the gap between the first and third worlds is too great to be bridged. Most third world countries are nowhere near even starting basic work in biology, leave alone learning the more advanced techniques. Their position in respect of access to information and knowledge remains highly unfavourable given the highly iniquitous patent and intellectual property protection regime that exists today (Bidwai 1987).

Another commentator has similarly criticized the "spirit of imposing technocentric, dependence-producing and very expensive programmes on the peoples of the south." He continues,

> There appears to be a method in the brashness with which ill-conceived, unimaginative and ill-designed programmes by western/international health czars have been imposed on the peoples of the south. There are massive media assaults; there is co-option of high-profile intellectual lightweights of the south; there is diversion of attention from the commitments made at Alma Ata; there is creation of a market for goods and services from the affluent countries; there is the creation of dependence of the south on the north; and, above everything else, there is political con-

trol over the victims by the international syndicate—"follow the party line, or else we withdraw our assistance" in true World Bank/IMF style (Banerji 1990).

Such suspicions are held not only by isolated Luddites. The former Director-General of the World Health Organization has voiced generally similar, if more muted, views:

> Technology for the sake of technology is a dangerous addiction-producing drug. We must always bear in mind the practical implication of existing and new scientific knowledge for the benefit of the masses of the world's population (Mahler 1976; quoted in Parker et al. 1977).

*Attitudes toward genetically engineered living vaccines.* It is difficult to overestimate the political environment surrounding the release of recombinant organisms, particularly in countries other than those in which the living vaccines were developed. In July 1986, a small-scale rabies vaccine test was conducted by scientists from the United States and the Pan American Health Organization at a rural facility belonging to the Pan American Zoonoses Center in Azul, Argentina. The test, conducted on 20 cows, involved a *Vaccinia* virus into which the gene for a rabies envelope glycoprotein was introduced by genetic engineering (see Appendix A for methodology). The purpose was to verify passive transmission of the *Vaccinia* virus from inoculated to control animals that were housed together. As it happened, the inoculated animals developed anti-rabies antibodies and the control animals did not. Although there is some controversy concerning the Argentine government's prior knowledge of the experiment, information about the test caused a furor both in Argentina and among Americans opposed to testing of genetically engineered organisms (Joyce 1986; Fox 1987a, b).

*The political significance of population-based studies.* In an article titled, "Indo–US Vaccine Project Worries Scientists," the Indian Express newspaper (August 17, 1987) said:

> The concern is about the enormous epidemiological data that will be collected as part of the vaccine trials. Samples of blood, sera and cells can tell a lot about the genetic make-up of a population, its immunity and antibody profile—collectively known as the "herd structure." . . . Because of its potential uses to biological warfare specialists no country gives its epidemiological data.

While these opinions may not be widely held, it is prudent to be aware of their existence. Despite the naysayers, one may be confident that safe, efficacious, and cost-effective innovations will become accepted in the field and in the marketplace.

# 3

# Epidemiology and Ethics in Transferring Technology

Any trial to evaluate the safety and efficacy of biomedical innovations has two essential supports. The first is *epidemiology,* which explores the determinants and distribution of diseases in human populations and defines the rules for obtaining and analyzing data. The other is *ethics,* which delineates the ways in which information may be appropriately obtained from participating human subjects. This chapter provides a background in these two areas; chapter 4 will describe the actual conduct of clinical trials.

## Epidemiology

The evolution of an innovation from creator to user is like a relay race, in which the baton is passed from one participant to another. At each stage, success depends on a smooth handoff to a trained, capable, and motivated colleague who already has a running start. Lack of preparation, planning, or teamwork are handicaps that waste everyone's efforts and resources.

In the same spirit, the biotechnologist must at some point relinquish his or her creation to the next professional, to continue the sequence of development, evaluation, and implementation. When field trials are indicated, the time has come to call the epidemiologist.

### Uses of Epidemiology

The methods of epidemiology provide essential descriptive information about the patterns of transmission of the agents of disease. Epidemiologic principles also represent a critical tool to guide interventions and to define the benchmarks by which these are assessed.

The astute observations of numerous investigators have established causal linkages for many human illnesses. Descriptive epidemiology is practiced widely in industrialized countries where chronic diseases are major causes of morbidity and mortality. Biomedically oriented researchers, in collaboration with statisticians, generate and analyze data about associations between the occurrence of diseases and various environmental, dietary, and behavioral factors. Logic dictates that modification of the risk factors thus identified can alter the onset or severity of the disease.

In the developing countries, whose generally young, rapidly expanding populations are vulnerable to multiple infectious agents, epidemiologists cannot afford to be passive observers. Where pathogens are ubiquitous, resources minimal, nutrition marginal, ignorance widespread, and disease the regular consequence, there is little to be gained in searching for statistically significant risk factors. The biomedical innovator who intends to reduce the risk of illness in individuals, or to control a specific disease in the community, will find that the roots of most ill health are readily apparent.

*Strengthening epidemiologic competence in developing countries.* Individuals with skills in epidemiologic planning and analysis are now available in many countries. Field Epidemiology Training Programs (FETPs) have been set up in Indonesia, Mexico, Peru, the Philippines, Taiwan, Thailand, and Saudi Arabia, with assistance from the Centers for Disease Control and Prevention (CDC; U.S. Public Health Service), and support from the World Health Organization. Their programs are modeled after the Epidemic Intelligence Service (EIS), a 2-year training program in applied epidemiology that has been conducted at the CDC since 1951. The EIS has graduated more than 1700 professionals who are serving all over the world. By the fall of 1989, the FETPs had 111 graduates, of whom 98% had remained in their respective countries, having investigated 638 disease outbreaks and conducted 328 research projects (Music and Schultz, 1990).

Another specific epidemiology training program is the International Clinical Epidemiology Network (INCLEN), founded in 1982 and sponsored primarily by the Rockefeller Foundation of New York. By 1991 Clinical Epidemiology Units (CEUs) had been established in 26 medical schools in Asia (including India), Africa, and Latin America. "The role of these units is to promote a rational approach to clinical and health care decision making, drawing on the methods of clinical epidemiology, biostatistics, health economics and health social science" (Halstead et al. 1991). All CEUs have established their own training courses and workshops, and serve as self-sustaining training nuclei within their regions.

Epidemiologic collaboration and training in developing countries are an integral part of many other programs, including those of the World and Pan American Health Organizations (See chapter 6).

## The Significance of Disease in the Population

*Data on the population.* Before the amount of disease in a particular region can be measured, certain baseline demographic information must be known about the resident population. The data are gathered in three major ways: by *enumeration* (in a census), by *continuous registration* of vital events, and through sampling surveys.

Good census data are invaluable for planning a valid representative population sample for randomized controlled trials in the field. A properly designed census provides data about the number of residents in any particular region, their age and sex distribution, place and type of residence, occupation, level of education and income, ethnicity, language, migration history, and a host of other characteristics. These descriptors are essential to understand the occurrence of disease in diverse population segments, to plan programs for control, and to evaluate the effectiveness of interventions.

Unfortunately, recent and reliable census figures are unavailable for many areas in which field trials or implementation programs may be contemplated. In such places, continuing registration of vital events, while perhaps legally mandatory, is incomplete, and investigators may need to generate the necessary data before a field trial can begin. The cost, in time and resources, is substantial, but there is no reasonable alternative if scientific validity is to be achieved and regulatory requirements for licensing and distribution are to be met.

## Sample Surveys

Except in unusual circumstances, field trials of biomedical interventions are conducted not on entire populations but on appropriately drawn samples. Defined as the process of extracting a representative set of examples from a larger population, correct sampling methodology is crucial to the reliability of almost all field projects.

*Sample size and cost of the study.* The number of individuals needed in the sample depends on the incidence of the disease in question, the presumed protective efficacy of the vaccine, anticipated frequency of adverse effects, the required statistical confidence level and power, the particular study design, and the nature of the materials being evaluated.

The "classical" clinical trial involves a pool of susceptible individuals. Each subject receives either the vaccine or an inert placebo, allocated by a strict process of "blinded" randomization. That is, the material received is not known to the subjects nor to the investigator until the trial is completed and the assignment code is revealed. The trials of the new inactivated polio vaccine in the 1950s, and of the first measles vaccine in the 1960s, were of this type. At that time there was no other available means of prevention of those diseases.

Once a vaccine has been developed and is found safe and efficacious against a particular disease, the world has changed forever. A simple placebo-controlled trial is no longer acceptable. Once a vaccine is available, subjects cannot be exposed to the risk of disease just to see if another vaccine might be in some way better. In that case, a trial could randomize subjects to receive (1) the new vaccine, or (2) the existing vaccine, rather than an inerto placebo. Such a procedure is rational, ethical, and very expensive. Assume that the existing vaccine prevents disease in 90% of those who receive it. Then, *ceteris paribus,* the pool of subjects must be ten times larger to yield the number of cases that would have occurred in the same population when unprotected. Cost and complexity would increase proportionately. A typical sample size calculation is shown in Table 3–1.

Despite the existence of a widely used and efficacious vaccine, it may still be possible to conduct a placebo-controlled trial. Such an exception occurred when the Swedish government decided in 1979 to terminate the routine use of whole-cell pertussis vaccine in infants because of its reputed adverse effects. In 1981, new acellular pertussis vaccines were licensed for routine use in Japan, reportedly with good results. Accordingly, a field trial was set up in Sweden in which two Japanese acellular vaccines were field tested against a placebo, with the approval of ethics committees from Swedish and international organizations. This trial was carried out in 1986 and 1987 (Blackwelder et al. 1991).

If a new vaccine is developed whose primary advantage is greater safety; i.e., reduced incidence of adverse effects, rather than greater efficacy, then the required number of subjects would increase in proportion to the rarity of the adverse effects, and the trial might become financially and logistically unfeasible. In that case the vaccine could be licensed and introduced for regular use, with some sort of monitoring and surveillance system to detect and report adverse events among the large pool of routine recipients.

*Sampling procedure.* A "sampling frame" such as a list of electors, patient register, school roster, or map may be used from which to select sample members by a predetermined process of randomization. Sampling elements (units) may be villages in a province, households in a community, mothers of children under 5, infants born between last January and May, or whatever is appropriate to the purpose of the study. Sample size is not the sole determining factor: a smaller sample that is truly representative of the universe being studied is more valid than

TABLE 3–1.   Sample Size Needed for a Clinical Trial or Cohort
Study of Pneumococcal Vaccine

| Annual Number of Cases of Systemic Pneumococcal Infection per Population at Risk[a] | Sample Size if Desired Protective Efficacy to be Detected Is | |
|---|---|---|
| | 50% | 75% |
| 8.5 per 100,000 | 1,417,400 | 466,060 |
| 25 per 100,000 | 481,858 | 158,446 |
| 500 per 100,000 | 24,010 | 7,902 |

*Source:* Clemens and Shapiro 1984.

[a]Sample size is the total number of subjects, with equal numbers of vaccinees and controls, for detection of the cited efficacy at $P<.05$ and with .90 power in a study with one-year follow-up of patients. An annual incidence of 8.5 per 100,000 corresponds to the incidence in the general population; an annual incidence of 25 per 100,000 approximates the incidence in the elderly population; and an annual incidence of 500 per 100,000 is an arbitrarily high figure chosen to illustrate sample-size requirements for patients at extremely high risk.

a larger but poorly drawn sample to determine the generalizability and applicability of the findings.

It is often believed that simple random sampling, with an equal risk of selection of any unit in the population, will meet the criterion of representativeness. Ideally this is so if all relevant subsets of the larger population are present in sufficient numbers within the sample. However, it may be necessary to draw a very large sample to assure such representativeness, the more so when the underlying population is diverse; and the cost involved may be beyond the means of the project. For this reason techniques may be applied in which the risk of selection is not the same for all elements in the population.

Consider for example a community of 5,000 inhabitants in which a 10% sample is to be taken in a study of filariasis infection, with a budget just sufficient for that number. In this village 3%, or 150 people, belong to a poor minority ethnic group. A random 10% sample would capture about 15 people from this group, which may be insufficient for statistical analysis of the influence of ethnicity on infection. Accordingly, the number of people chosen from the minority group may be increased, say to 50, and the number from the majority population correspondingly reduced from 485 to 450. The total sample size and sampling cost are unchanged, but the minority sample of 50 is sufficient for statistical validity in this particular context. Subsequently the survey results from the majority population can be corrected if necessary to produce an adjusted sample representative of the entire community. With analogous reasoning the total population can be stratified into subgroups by age, residence (e.g., distance from a point source of infection), or other characteristics to assure adequate numbers of each subgroup within the sample. Sampling may then be randomized within each stratum (also called blocked randomization). The same process can reduce the effects of confounding variables.

The main difficulty with these methods is that the relevant characteristics on which the population is to be stratified must be known in advance, and adequate demographic information must be available by which the total population can be allocated into the various strata.

These methods may be used to obtain baseline data about disease distribution within populations of individuals, households, or communities. The same principles operate in relation to the conduct of randomized clinical trials, as discussed below.

*Cluster sampling.* Investigators always seek methods that are simple, inexpensive, and effective. The technique of cluster sampling, used by the Expanded Programme on Immunization (EPI) to assess immunization coverage, meets these criteria. Although it does not provide the degree of statistical precision of some other sampling formats, cluster sampling is considered satisfactory for the requirements of the EPI.

In practice, the method consists of

1. identification of the geographical areas of interest;
2. identification of the age groups of interest;

3. random selection of 30 sites ("clusters") from within each geographical area for which individual results are desired;
4. random selection of a starting point (household) within each site;
5. selection of seven individuals of the appropriate age from within each of the 30 sites.

At each site, selection begins in the starting household and continues to the next nearest household until a total of seven individuals is enrolled. All individuals of the appropriate age living in the last household falling into the sample are included, even if this means including eight to ten individuals in the cluster rather than the required minimum of seven (Henderson and Sundaresan, 1982). This technique can be adapted to other sample surveys.

*Required degree of precision.* Competing demands for limited resources mandate the most efficient means of obtaining data in developing countries. On the other hand, only statistically valid study protocols can provide the information needed for reliable conclusions and can satisfy regulatory agencies. There should be no conflict between the investigator's twin obligations to conduct studies that are both cost-effective and conclusive. Procedures should be kept as simple as is feasible. It is a mistake to confuse the needs of adequate and valid data gathering with the criteria of ultrasophisticated research.

Foster (1987) criticized many projects in developing countries on grounds of overelaboration:

> We find many instances in which ego-involvement has gotten the best of the researcher. This tendency seems particularly characteristic of those researchers who have had overseas postgraduate training. In their desire to demonstrate their competence, . . . they are primarily concerned to carry out the most elaborate project possible, quite forgetting the need for practical utility of their research results. . . . They overlook entirely the need to address the question of the ends for which the research is carried out. . . . The primary requirement for good operational research—keep it simple—is lost sight of in the researcher's desire to do a study that will excite peer admiration.

## Shortcomings of Sample Surveys

Weaknesses of survey research have been pointed out by many authors; for example:

> (a) People typically respond to survey questions with the first thing that comes to mind, and then become committed to their answer; (b) people typically provide an answer to any question that is posed, even when they have no opinion, when they do not understand the question, or when they hold inconsistent beliefs; (c) survey responses can be influenced by the order in which the questions are posed, by whether the emphasis is on speed or accuracy, by whether the question is closed or open, by whether respondents are asked for a verbal or numerical answer, by interviewer prompting, and by how the question is posed (Fischoff et al. 1980).

## Basic Measures of Disease Frequency

The primary measures of the frequency of a disease or condition in a population are *prevalence* and *incidence:* in practice these are often confused or interchanged.

Prevalence, in its simplest guise, is generally defined as the proportion of individuals in a defined population who have the condition or disease in question at a given point in time (point prevalence) or at any time during a specified interval of time (period prevalence). It is a ratio of cases to noncases.

The incidence rate considers the number of new cases of a disease that occur per population unit (e.g., per thousand people) over a specified time period. It is a true rate (like kilometers per hour), because it states a frequency of occurrence in the numerator and a measure of time in the denominator. It is also a ratio of those individuals with onset to those without. For example,

$$\text{Annual incidence rate for occurrence of a specified condition} = \frac{\text{Number of new cases of the specified condition occurring in a defined population during a year}}{\text{Number in that population at midyear of the same year}} \times 10^n$$

where the exponent $n$ defines the population unit. Embedded in this deceptively simple formula are a number of methodological complications that must be understood if useful information is to come from a field trial.

Exactly what are we looking for? Is it infection with a specific agent (e.g., *Toxoplasma gondii*) or with any one, or perhaps several, of a group of related agents? For example, are we interested in any diarrhea or only in diarrhea caused by rotavirus, *Shigella, Salmonella,* or enteropathogenic *E. coli?* Are we interested in any pneumonia, or in pneumonia caused by *Haemophilus influenzae, Streptococcus pneumoniae,* certain viruses, and so forth? Can the infection be inapparent or subclinical? Are we doing molecular epidemiology, looking for specific strains, isotypes, serotypes, or other genetic variants of these pathogens?

The expression of a disease may vary under different conditions. For example, a case of measles may be defined as (1) generalized rash of 3 or more days' duration; (2) fever of 101F (38.3C) or more; and (3) cough, coryza, or conjunctivitis (Orenstein et al. 1984). However, the severity of clinical measles may be different in children after immunization (Aaby et al. 1986; Aaby 1992). Therefore a survey of clinical measles occurrence based on a case definition that does not consider previous immunization may have a built-in bias.

Having written an iron-clad case definition that is both inclusive (of all cases) and exclusive (of everything else), it will be necessary to determine for each individual whether those criteria are met. Using the definition given above, the methods of observation, use of a thermometer, and inquiring about duration of illness can ascertain whether a child does or does not have measles. To ascertain the etiology (causality) of most illnesses, more complicated resources are usually required.

For some diseases it is not easy either to define a "case" or to ascertain infection. In schistosomiasis, persons with light worm burdens typically show no clinical signs. Every individual enrolled in an antischistosomal vaccine trial would need to have costly periodic laboratory studies to determine whether infection had occurred since the previous examination. The appearance of schistosome eggs in the stool is the "gold standard" to demonstrate infection. But absence of eggs cannot rule out an infection with single-sex or sterile worms, inhibited oviposition, or blockage of egg passage to the exterior. A serologic test may be needed. Although the definition and ascertainment of schistosomiasis is particularly complex (Basch, 1993), any pathologic process will present its own peculiar difficulties.

The level of diagnostic services depends on the purpose and design of the epidemiologic study. One may wish to determine whether malnutrition in children under five exceeds, say, 5% of the population at risk, which might be the predetermined prevalence level to trigger a provincial control program. A relatively simple and inexpensive clinical survey, with some laboratory backup, may be sufficient to make a yes or no decision. It is quite another matter if a carefully controlled randomized clinical trial of a new vaccine is undertaken.

To determine the efficacy of a *community-based intervention* such as a water supply or vector control program, it is necessary to conduct at least two prevalence surveys: before and after. Where there are *personal interventions* such as administration of vaccines in a randomized format, individuals must be identifiable and traceable over the entire period of the trial. Depending on the circumstances, it may not be sufficient to verify the presence of infection or disease, but it may be necessary also to evaluate the level of prior susceptibility, for example, by determining pre-existing antibody titer, in each participant before allocation to vaccine or control groups.

### Error, Bias, and Comparability

Careful planning is needed to avoid errors in study design and in the collection or interpretation of data. *Random* or *sampling errors* occur, for example, in selecting inappropriate samples or in extrapolating information from such samples to larger populations. The simplest way to minimize random errors is to increase sample size or sampling frequency. Practical considerations of cost and logistics always require compromises.

*Systematic errors* that lead to *bias* may jeopardize the validity of epidemiologic studies. The basic tenets of the experimental method require that intervention and control groups be as similar as possible with the sole exception of the factor being studied. If individual subjects are unwittingly systematically included or excluded from one or the other category, the study suffers from *selection bias*. In addition to internal validity there is also external validity, or the degree to which the entire study cohort is representative of the total population to which the findings of the study are to be applied.

Unanticipated factors may always emerge. For example,

> Twins carry a high risk of mortality, particularly in the first year of life. In some societies they may represent as much as 15% of all infant deaths. At the same time the rate of twinning varies from one ethnic group to another, in West Africa this variation may be as much as threefold (Hall and Aaby 1990).

In a comparative study of infant mortality among African communities it would clearly be necessary to have information about rates of twinning in each locality, which may not be evident a priori to researchers writing the study protocol.

A second kind of bias often encountered is called *information bias,* which may take several forms. Subjects with different backgrounds may respond to a questionnaire differently depending on the specific wording of questions, or recall or interpret information differently in accordance with their cultural beliefs. Subjects who have lost a spouse to AIDS may be less likely to admit that cause in comparison with those whose spouse has died of a heart attack. Physicians may show similar tendencies in stating the cause of death on a death certificate.

Information bias extends to subjects who know, or think they know, whether they have received the vaccine or the placebo, and to investigators and study employees who may be influenced by some characteristics of subjects. In the one case, subjective perception of symptoms and articulation of complaints may be correlated with the subject's notion (or conviction) of which dose was received. In the other, interpretations of borderline antibody titers might tend one way or the other depending on the subject's name, place of residence, occupation, or almost any other factor that may influence the technician. Multiple blinding is mandatory, so that the category of individual subjects is unknown to themselves, to the experimenters, to laboratory personnel and to statistical analysts until the end of the trial.

A third kind of systematic error that is sometimes troublesome is *confounding bias.* The effect of the factor that one wishes to study may be confounded by some other circumstance that may or may not be recognized. For example, in an epidemiological study of filariasis we may want to know the effect of water supply on local transmission. As a start we can compare the number of mosquito bites per person-night in two otherwise similar villages, one with an old central open well from which people draw pails of water, and the other whose well has been replaced with a new piped water system. Although the villages appear otherwise identical to us, an enterprising salesman has given the residents of the newly plumbed village samples of a certain soap that leaves a perfume-like residue that inhibits mosquito bites. The other village continues to wash with well water alone. The water-supply-based analysis is destroyed by the confounding factor of soap in one village and not in the other.

While it is a simple matter to warn against these types of bias, it is much more difficult to assure their exclusion, particularly in the case of studies carried out in relatively remote and unfamiliar areas. A comprehensive list of 25 types of bias was presented by Last (1983).

## Principles of Diagnosis

Accurate diagnosis is the underpinning of most epidemiologic studies. Clinical trials of vaccines and drugs are held hostage to diagnostic procedures. Diagnosis leads to decisions about the status of illness in an individual, or of disease in a population. Diagnostic procedures are used also to identify pathogens in the environment (water, soil, or food), or in products such as blood for transfusion.

Every unambiguous diagnostic procedure results in a determination that may be qualitative (such as the presence or absence of malaria parasites in the blood), or quantitative (such as the titer, or concentration, of a particular antibody), or both.

Although the result of a diagnostic procedure may be unequivocal, it is not necessarily correct. There are four possible outcomes of any diagnostic test: two are correct (true positive and true negative), and two are incorrect (false positive and false negative) (see Figure 3–1).

The *sensitivity* of a test is a measure of the likelihood that the test was positive (not a false negative) in those in whom the condition is really present. If there are no false negatives, the test has a sensitivity of 100%.

Similarly, the test may classify some people as positive when in fact they are not. The proportion of such false positives defines the *specificity* of the test, which reaches 100% when no person without the condition is misclassified as having it. (Hint for the memory: the words false positive start with the letters f and p, which are also found in the word specificity; negative starts with n which is also in sensitivity.) The sensitivity and specificity of diagnostic procedures are always estimates, because the real truth is never known with certainty.

A test has a high *positive predictive value* when it is very likely that an individual determined to be positive actually has the condition. A test has a high *negative predictive value* if there is a high probability that a person classified as negative really does not have the condition. The actual predictive value of the test is a function not only of its sensitivity and specificity, but also of the prevalence of the condition in the population being studied.

*How predictive value changes with disease prevalence.* Assume that a diagnostic procedure is actually 99% sensitive and 99% specific (1% false positives and 1% false negatives), and that the *true* prevalence of a condition is 1 case per 100 persons in population A and 1 case per 1,000 persons in population B.

Figure 3–2 shows the outcomes if 100,000 randomly selected persons are tested in each population. Where the true prevalence is 1 per 100, half of the people found positive are misclassified. In the population in which the true prevalence is 1 per 1,000, fewer than 10% of the persons who test positive are actually positive. The positive predictive value of any test declines as the prevalence in the population diminishes. When true prevalence is zero, every positive test is incorrect. In the same manner, the negative predictive value of a procedure is zero if every person in a population has the condition.

FIGURE 3–1.    Possible outcomes of diagnostic tests.

|  |  | TRUE PREVALENCE[1] | |
|---|---|---|---|
|  |  | Positive | Negative |
| TEST RESULT | Positive | True Positive | False Positive |
|  | Negative | False Negative | True Negative |

[1] Which can never be known with certainty.

FIGURE 3–2. How predictive value changes with disease prevalence. Adapted from Basch. 1990.

Case A. True prevalence = 1 case/100

TRUE PREVALENCE

|  | Positive | Negative | Total |
|---|---|---|---|
| Test Positive | 990 | 990 | 1,980 |
| Test Negative | 10 | 98,010 | 98,020 |
| Total | 1,000 | 99,000 | 100,000 |

Case B. True prevalence = 1 case/1000

TRUE PREVALENCE

|  | Positive | Negative | Total |
|---|---|---|---|
| Test Positive | 99 | 999 | 1,098 |
| Test Negative | 1 | 98,901 | 98,902 |
| Total | 100 | 99,000 | 100,000 |

The prevalence of a disease also influences the strategy for application of measures such as vaccination. A very rare condition would not warrant universal immunization even with an excellent vaccine, and decisions must be made on a case-by-case basis about vaccine strategies for conditions of intermediate prevalence. An example for use of hepatitis B vaccine is shown in Table 3–2.

## Seroepidemiology

The mainstay for vaccine efficacy trials, and a key technique for studies of infectious disease epidemiology in developing countries, is the determination of specific serum antibody titers (see Appendix A). An obvious drawback of all serologic work is the need to take blood from the subject. Few people are pleased to provide even one drop taken by capillary tube or filter paper from the earlobe, heel, or finger. Many people in developing countries believe that the total amount of blood in the body is fixed and limited, and that any quantity removed will never be replaced. Some associate blood with other body fluids, and believe that removal of blood will, for example, diminish the production of semen. In such

TABLE 3–2.   How Hepatitis B Vaccine Strategy Varies with Prevalence

| Factor | Prevalence Patterns | | |
|--------|-----|--------------|------|
|        | Low | Intermediate | High |
| HBs antigen prevalence | 0.2–0.5% | 2–7% | 8–20% |
| Anti-HBs prevalence | 4–6% | 20–55% | 70–95% |
| Childhood infection | infrequent | frequent | very frequent |
| Neonatal infection | | frequent | very frequent |

*Recommendations for Immunization*

| Low Prevalence | | Intermediate to High Prevalence | |
|----------------|--------------|------------|-------------|
| Preexposure | Postexposure | Preexposure | Postexposure |
| high-risk groups | accidental exposure | all infants | infants of positive |
| health care staff | sexual contacts | | mothers |
| dialysis patients | infants of positive | | |
| drug addicts | mothers | | |
| military recruits | carriers | | |
| male homosexuals | | | |

*Source:* Adapted from Zuckerman 1985.

circumstances the taking of larger volumes (several milliliters) of blood by veni-puncture is strongly resisted, limiting the number and range of observations that can be made. Lymphocyte mitogenesis or similar studies of cell-mediated immunity, for example, are almost impossible in the absence of an adequate number of blood cells. Dried blood on filter paper is not suitable for some immunologic tests nor for isolation of viruses or organisms (Lamm, 1988). Another increasingly significant drawback of serologic work is the risk to laboratory personnel in handling sera possibly contaminated with hepatitis B, human immunodeficiency virus, or other pathogens.

Because of the problems inherent in obtaining and processing blood, some workers have sought noninvasive ways to determine antibody levels, such as in saliva or urine. Although a number of such studies have been published, these methods are generally not considered adequate replacements for serology. On the other hand, antigens of certain pathogens can be detected in urine and, depending on the study, might provide a useful indication of present infection and a check on the efficacy of immunization or chemotherapy.

Serologic testing finds two main applications in epidemiology: to establish seroprevalence, and to follow seroconversion (Orenstein et al., 1988).

*Seroprevalence* is a determination of the titer of the antibody of interest in a population at a point in time. A seroprevalence survey can be done as a screening procedure prior to selecting the study cohort for a vaccine trial, since there is usually little to be learned by vaccinating an already immune person. It is important to recognize, however, that the presence of antibody may or may not correlate with *protection* from infection or disease. Similarly, the absence of detectable circulating antibody does not necessarily prove that an individual is susceptible. Antibody levels may wane while cellular immunologic memory remains intact. Although it is the best tool available, seroepidemiology provides only an approximation of the level of susceptibility or immunity to specific pathogens.

Advantages of a seroprevalence study are:

1. It provides baseline data on previous population exposure to the antigen in question.
2. The data can serve double duty as the preprocedure titers in a seroconversion study.
3. It helps in selecting the proper age for immunization by identifying the age at which a substantial number of individuals in the population already have naturally acquired antibodies.

However, this information might be misleading after epidemics, in the case of diseases (e.g., pertussis) that normally have 3- or 4-year cycles of high and low prevalence, or where recent unusual environmental conditions, such as drought or flood, have affected the transmission of the pathogen in question.

It may be mentioned also that *nonspecific immunity* stimulated by BCG or by some kinds of vaccine adjuvants (see Appendix B) may be partially protective against a wide variety of antigens. Therefore, in a vaccine-versus-placebo protocol for a randomized controlled trial, the placebo must be carefully defined. The placebo may be the complete vaccine minus antigen, or the vaccine minus antigen and adjuvant, or some other formulation, with potentially different outcomes for each.

In a population in which immunizations have been given in the past, a seroprevalence survey may not be able to distinguish between antibodies induced by vaccines and those resulting from infection. In the United States, BCG vaccine is not used routinely because everyone vaccinated would then have a positive tuberculosis skin test.

*Seroconversion* studies require at least two blood samples: the first taken before or at the time of immunization (or exposure); the second, some time later. A rise in antibody titer to some predetermined level is considered a criterion of seroconversion.

## Herd Immunity

The mathematical modeling of transmission patterns of infectious agents has long been a standard exercise in epidemiology classes. These models generally employ a freely intermixing population with one or more index cases and a uniform probability of contact between them and all remaining susceptible individuals. The basic reproductive rate of a pathogen transmitted from person to person is defined as the average number of secondary cases resulting from a primary case. If each case produces on average one other case, then the infection will remain stable and endemic in the community. If each case produces more than one additional case an epidemic will result; if less, then the infection will eventually die out in a closed community but may continue where births constitute a continuing pool of new susceptibles, or where new cases may be imported through visitors or immigration.

In fact, humans do not behave like ping pong balls in a box, and pathogens are not transmitted with equal likelihood to every uninfected individual. The success of the smallpox eradication program can be attributed to this realization and to a change of strategy in which the original concept of simply vaccinating every person in the world was replaced by a focal one of surveillance and containment, based on aggressive casefinding and the tracing and vaccination of contacts.

Within these non-panmictic populations, many factors determine the probability of association between any two humans. The behavioral and cultural patterns of their parents determine the exposures of infants and young children. In developing countries the generally warm climate and high population density favor certain kinds of transmission, and the periodic or episodic concentration of young children can favor the transmission of pathogens. Where children are brought to markets or social gatherings, or to educational or public health functions such as motherhood classes and immunization clinics, particularly mass campaigns where large numbers of people assemble from diverse localities, the probability is high for interchange of pathogens. Measles is well suited to such transmission, as infectious viruses may remain in the atmosphere for many minutes even after the index case has left the scene.

An increase in the proportion of immunes and a reduction in the proportion of susceptibles in the population will result in a diminution of transmission so that the remaining susceptibles are less likely to encounter the pathogen by contact with an infected person. This phenomenon is known as *herd immunity,* by which nonimmunized individuals in endemic areas or highly immunized populations are protected to a substantial extent from diphtheria, pertussis, measles, and other childhood diseases. Conversion of individuals from susceptible to immune through immunization rather than through natural disease also eliminates those pathogenic organisms that would have been transmitted by that person had he or she become naturally infected and infectious.

In addition to the benefit of herd immunity, attenuated live oral poliomyelitis virus vaccine can be transmitted from the vaccines to other people, thereby protecting individuals who have not themselves received the vaccine, and augmenting the reach of an immunization program.

*Erroneous assumptions regarding herd immunity.* One might think that the best selfish tactic is to refuse immunization for oneself to avoid adverse effects, but advocate it for others in the hope of inducing herd immunity, thereby lessening the risk of illness. Although rational, this plan is certainly not foolproof because any susceptible individual, however isolated, remains exposed to infection. Many instances are known of cases, or even of epidemics, of disease within presumably well-immunized populations. Possible explanations for such outbreaks include:

1. The use of impotent vaccine; e.g., inadequate manufacturing or testing, outdated stocks, or a break in the cold chain resulting in exposure to heat and inactivation of live virus.

2. The existence or introduction of variant forms of the target pathogen, some of which are not affected by the vaccine-induced protection (e.g., influenza).
3. The presence of subgroups with low vaccine coverage within larger populations. These pockets may be difficult to reach because of geographic or cultural isolation, or they may consist of groups with religious or other objections to immunization.

An outbreak of poliomyelitis on Taiwan was described by Kim-Farley et al. (1984). Between 1975 and the beginning of 1982 Taiwan had been free of outbreaks of paralytic polio, with fewer than 9 sporadic cases reported annually. Between May and September, 1982, 1,031 cases of type 1 paralytic polio (median age, 16 months) were reported to authorities. Approximately 400,000 births per year were registered at the end of 1980. An estimated 80% of infants had received two doses of trivalent oral polio vaccine, whose efficacy was calculated at 82%, 96%, and 98% after 1, 2, or 3 doses, respectively. Despite this degree of coverage, the outbreak occurred primarily in the unvaccinated children, who were estimated to have 80 times the risk of paralytic polio of a fully immunized child.

Similar situations have occurred in the 1992 outbreak of polio in the Netherlands, referred to earlier, and in a number of epidemics of measles, rubella, and other viral diseases that have been limited to small groups of people who have refused immunization for religious reasons

## Disease Surveillance

Surveillance is an ongoing process by which the occurrence of diseases or events is monitored within a particular geographic area and period of time. Without adequate surveillance there may be little or no basis for selecting one site over another for a field trial, or to evaluate the consequences of an intervention in terms of adverse events, or a change in the incidence of a particular disease. In his excellent review of epidemiologic surveillance in developing countries Frerichs (1991) emphasized the resources that must be devoted to a good surveillance system and the need to justify its components.

*Passive surveillance* usually depends on the study of clinical records or data routinely collected at health centers, clinics, or hospitals; or on self-reporting in various ways by individuals. The data collected must be clearly defined and used in appropriate ways to improve the efficiency of the system and support health goals. Consider this account from Myanmar (Burma) of a surveillance system gone wild:

> Peripheral-level health workers were expected to complete more than 30 sets of forms either on a daily or monthly basis. At the rural health center level, the local staff had to submit a set of forms with 1160 variables every month. Of these variables, 72% were requested by the Division of Disease Control, the unit responsible for epidemiologic surveillance. At the next administrative level, Township medical officers each month had to process and review these variables, plus 786 more variables on township-level activities, for a total of 1946 variables. . . . By the time the information was received at the national level, processed and analyzed,

and included in a detailed report, three years would pass. The information was then too dated to be used for decision-making. Although the surveillance report represented the time and effort of countless health workers, no one trusted the data (Frerichs 1991).

*Active surveillance* or casefinding may be done in schools or markets, by door-to-door canvassing or by similar means. During the smallpox eradication campaign great efforts were expended on active surveillance, including cash incentives to the public for reporting a suspected case.

*Sentinel surveillance* has several meanings. One is the observation of subjects to detect the occurrence of infection. Sentinels may be non-human; for instance, a chicken tethered to a forest platform to pick up mosquito-borne viruses, or mice floated in a cage near a transmission site to detect the presence of infective schistosome cercariae. Obviously the species and conditions must be appropriate to answer the particular question under study. In another meaning, sentinel surveillance refers to the scrutiny of records at particular "sentinel" clinics or health posts located in certain critical sites to pick up early cases of disease.

## Types of Epidemiologic Study Formats

Over the decades several basic plans for studies have been defined by epidemiologists. The nonexperimental or observational formats are cohort studies and case-control studies, while the experimental approach is through randomized controlled trials.

*Cohort studies.* A cohort is simply a defined set of people. Studies on cohorts can be carried out in various ways. In a *prospective cohort study,* persons enter the study at a particular point in time and are followed into the future with continual or periodic observations about the occurrence of events of interest, such as certain illnesses. For example, the cohort may consist of members of a certain group having one or more defined characteristics, such as having or having not been immunized with a certain antigen or abstaining from the consumption of alcohol. *Retrospective cohort studies* are similar but start with the present and look at the health history of individuals in groups containing members who did or did not have a certain characteristic, behavior, or intervention in the past. Clemens and Shapiro (1984) studied patients with pneumococcal pneumonia. They looked at the proportion of cases in vaccinated and unvaccinated patients caused by the specific pneumococcal serotypes contained in a vaccine, to determine whether the vaccine conferred cross-protection against other serotypes.

*Case-control studies.* This type of study is used to compare two groups, cases and controls, with respect to the frequency of their past exposure to a particular factor. In a classical case-control format a group of women with lung cancer, and another group of similar ("matched") women without lung cancer, are questioned about their previous cigarette smoking habits. A certain proportion of each group will admit to the exposure, and it is then a simple matter to calculate the ratio of the odds of smoking among cases to the odds of smoking among non-cases, or simply, the *odds ratio*. If 80% of lung cancer patients, but only 16% of non-patients, had been smokers, then the odds ratio of 5:1 would raise suspicion of an association and trigger further investigations. An *attributable risk* can then be calculated as an estimate of the impact of that particular factor (e.g., cigarette smoking) on the causation of the resultant effect (cancer of the lung).

Identical kinds of studies can be done to estimate the efficacy of various health care services, including immunizations. For example, a large group of persons with tuberculosis and a TB-free control group may be questioned about their BCG immunization status (Houston et al. 1990, Comstock 1990). A null hypothesis would say that there should be no difference in the proportion of each group that had ever been immunized with BCG. If it should be found that a significantly smaller proportion of TB patients had been immunized in comparison with the controls, then appropriate statistical procedures should be performed to evaluate the degree of association.

Advantages of the case-control format are that it is fast, simple, and relatively inexpensive. "Even for the evaluation of vaccines against acute infectious diseases, the economy and feasibility of observational studies make them appealing" (Comstock 1990). The disadvantages relate primarily to the ease with which all of the types of biases can be incorporated. The matching of controls must be done very carefully according to clearly predetermined criteria depending on the particular study. Information bias can be troublesome because the study depends on recall by persons who may have some incentive to over- or underestimate exposure (for example sexual practices), and on interviewers who are likely to know which participants are cases and which are controls. Although some careless and poorly planned case control studies have been reported, when properly done this procedure can develop useful epidemiological information.

*Randomized controlled trials.* This general format describes experiments in which future outcomes in a group of individuals who receive an intervention of some kind are compared with outcomes in one or more comparable groups of people who do not receive the intervention. Randomized *clinical trials,* which evaluate the efficacy of various forms of treatment, use patients as subjects (the ambiguous abbreviation RCT sometimes causes confusion). *Field trials* typically determine the safety and protective effect of a particular intervention among individuals who are not patients. Field trials may involve a few dozen volunteers or tens of thousands of individuals and represent the foremost methodology by which the products of biotechnology are evaluated.

The key element in all randomized controlled trials is the random allocation of subjects to the intervention and control groups. The procedure whereby this is done is carefully predetermined so that neither the investigators nor the participants can determine the group to which any individual subject is assigned. Randomization helps to assure that any observed difference in outcome between the two groups is attributable to the different interventions that they received, with the effects of chance diminishing as the sample size increases. It would be important, for example in an immunization trial in infants, that the characteristics of subjects assigned to vaccine and placebo (or alternative vaccine) groups not be greatly different with respect to age, maternal education, presumed exposure to a pathogen by place of residence, breastfeeding or bottlefeeding, or similar characteristics thought a priori to be associated with particular outcomes.

To help assure comparability between two groups, the total study population may be stratified on descriptors such as those listed above and then randomized from within each stratum to intervention and control groups. Postrandomization statistical methods are available to test the comparability of the groups thus created. The confounding effect of that variable is therefore eliminated within its stratum, and all the stratum-specific estimates of effect can be combined using advanced statistical weighting methods to develop an overall conclusion as to the efficacy of the intervention.

The main advantage of randomized controlled trials is the high level of credibility of the conclusions reached when the trials are properly designed and conducted. Disadvantages include complexity and cost as well as the time required to reach a conclusion. The need to obtain the extended collaboration of many individuals, as well as certain potential ethical issues, also may make randomized controlled trials difficult. See chapter 5 for further discussion.

Another type of randomized controlled trial of potential interest is the community-based intervention trial, in which a procedure, such as vector control, water treatment, or a radio or television educational campaign, is applied to a large population, a particular community, or a subset of a community, such as a factory, school, or church. Large evaluations for safety and efficacy of biotechnological innovations could include aspects of both field and community-based intervention trials.

## Epidemiologic Standards

A useful compilation of established epidemiological standards has been offered by Paffenbarger (1988), who commented,

> Since the conclusions and applications of epidemiological studies must rely so heavily on the values of circumstantial evidence, it is important that the procedures be designed to meet rigorous standards and principles that will ensure that the circumstantial evidence obtained will be as strong and convincing as possible.

Paffenbarger's ten formal standards are:

1. *Statistical association.* There must be a statistically significant positive or negative correlation between the presumed cause and the observed effect. . . . The presumption of cause and effect relationship must be plausible and logical in terms of common sense, and it should reflect the current understanding of the physiological and pathophysiological circumstances involved. . . . [However] no degree of statistical sophistication or immensity of population will compensate for faultiness in definitions, diagnostic criteria, study design, or data collection.

2. *Temporal sequence.* It must be shown that the assessed potential cause has preceded the assessed effect and has occurred within or over an accepted interval to be commensurate with a presumed induction (incubation) period.

3. *Consistency.* The findings . . . must be consistent in terms of demographic characteristics: age, race, sex, occupation, socioeconomic status, geography, etc. . . . The study plan must pay heed to procedures such as stratification by age, case-control matching, group homogeneity determination, and other influential elements of methodology. The findings should then display a corresponding consistency supportive of their acceptance as valid evidence of an observed association.

4. *Persistence.* The findings should be persistent . . . during successive intervals of time . . . not a momentary "snapshot" finding but a continuing relationship . . . demonstrated again and again.

5. *Independence.* The cause-and-effect relationship is seen both in the presence and in the absence of other influences . . . to clear away the problems of confounders that otherwise would tend to weaken the evidence and conclusions.

6. *Dose-response relationship.* With increased exposure to a presumed cause, we would expect a corresponding change in the risk or rate of [the effect].

7. *Specificity.* For best evidence we should find that the stated cause is specific for the observed effect. Frequently there is no association found between the suspected cause and a different effect.

8. *Alterability.* Here modification of the presumed cause produces a corresponding change in the effect . . . that is, in essence, an application of the dose-response relationship as noted above and might be taken to confirm it.

9. *Repeatability.* Different investigators in different places at different times and using different methods on different study populations should tend to come up with similar results if the relationship is strong and of central importance.

10. *Confirmation.* Epidemiological findings for physical activity seldom can seek experimental verification among human subjects in a clinic or laboratory, but some confirmations have been obtained from experimental studies.

## Other Nonexperimental Means of Assessment

*Models.* Three major aims of mathematical modeling in epidemiology have been identified by Anderson (1988):

1. to improve scientific understanding and precision in the expression of current theories and concepts;

2. to identify areas in which better epidemiological data are required to refine prediction and improve understanding;

3. to generate predictions both of a qualitative and quantitative character.

Modeling provides a way to estimate the potential impact of a product or methodology by conducting a preliminary evaluation on a computer. One major advantage of mathematical models is avoidance of the need to expend large amounts of time and money for real-life assessments in the field. A great deal of data can be incorporated into mathematical models, which can be designed to represent certain well-defined sets of conditions. However, the value of such an exercise can be no greater than the validity of the simulation, and in most cases it seems questionable whether the intuitive conclusions of a knowledgeable observer can be very much improved upon by the invocation of a simple model.

Among the many shortcomings of models are:

1. They tend to be too mathematical and theoretical.
2. Although a substantial amount of money may be expended in programmer salaries and computer time, models are generally too simple in relation to the real world.
3. They neglect nonquantitative elements such as cultural, environmental, and political factors.
4. They tend to be cross-sectional rather than longitudinal and therefore do not readily reflect the time element of changing conditions.
5. They commonly neglect cost considerations.
6. It may be impossible to determine the validity of the model without doing the field work anyway.

Despite these objections, modeling may be cost-effective in certain repetitive and well-defined situations, and may be the only available means of evaluation, e.g., where resources do not permit full-blown field trials.

*The Delphi technique.* The Delphi technique was developed during the 1960s by investigators at the Rand Corporation to enlist expert opinion in making predictions regarding national defense for the United States. As an instrument in part for technology assessment, more particularly for forecasting, the Delphi technique provides certain advantages and disadvantages, in comparison with actual field trials. The intention of Delphi studies is attainment of a reasonable consensus by an objective and systematic probing of the opinions of experts in the subject. In practice, written responses to carefully specified questions are obtained from a panel of experts who are anonymous to each other. The investigator then generates systematic modifications from tallies of the initial responses and sends these back to the panel, with one or more additional rounds or iterations of question and response until a group consensus is reached by aggregation.

Applications of the Delphi technique in health sciences have included strategies for drug treatment of infectious diseases, specification of needed characteristics of technical personnel, evaluations of problems and challenges in infectious disease epidemiology. Many recent Delphi studies have dealt with nursing policies. About two dozen studies in the health sector are conducted each year using the Delphi technique.

A typical Delphi study in technology forecasting was described by Bright (1978). The object was to predict the date of occurrence of certain possible technologic developments. In the first round the panel is asked to predict the date by which each specified event will occur. The director then collects and analyzes responses, including *nevers,* and returns certain data to each member of the panel. In the second round, each panelist reviews his or her predictions against the statistics of the group and may reconsider responses or provide reasons for nonconformity. In the third round the study director provides the summary of round 2 to all panelists, including explanations of the nonconformists, and panelists can consider the responses and provide counterarguments to the nonconforming opinions. A fourth and final round recirculates data, opinions, and counter-opinions and obtains the concluding opinion of all panelists, which constitutes the product of the study.

The Delphi technique has several advantages, including:

1. The process may be relatively simple and inexpensive compared with actual field trials.
2. Geographically isolated participants with different backgrounds and experience can contribute to the group response.
3. The process encourages the introduction and consideration of factors, including unpopular opinions, that any single investigator or group might have overlooked or neglected.
4. The ranking among procedural alternatives can forestall a possibly costly and time-consuming field trial considered unlikely to be successful.
5. The panel can contribute feedback to the laboratory by describing the most desirable characteristics of products such as vaccines, which can help to stimulate investigators, *inter alia*, to develop products with the specified properties.

Among the disadvantages of the Delphi technique are:

1. No actual field data are collected.
2. There is no guarantee that the panelists are not utterly wrong in their predictions.
3. A meritorious but untried procedure may be unfairly discouraged, and an innovative but useless one could be promoted.

## Quality of Evidence

Relevant decision makers must at some point determine whether a procedure should or should not be implemented, as described in greater detail in chapter 4. Both subjective ("political") and objective ("scientific") criteria enter, legitimately, into all such decisions. There is little to say in principle about political decisions, which may be made on transient and idiosyncratic grounds. On the other hand, objectivity demands evidence, but not all evidence is of equal value. An attempt to quantify the relative quality of different types of evidence was made by the Canadian Task Force on the Periodic Health Examination (1979), which established four weighted grades for health interventions, according to the quality of the evidence obtained about their effectiveness. The adaptation to epidemiologic analysis found in Table 3–3 is based loosely on the Task Force categories:

# The Ethical Issues

It is often assumed that a given biomedical innovation will benefit the residents of most developing countries. This may or may not be the case, in part because of complex and potentially troublesome ethical issues such as:

1. asymmetric power relationships (money, knowledge, capability, prestige) among the various actors;
2. cultural or religious differences in moral teachings, the meaning of justice, and societal goals; perceptions of individual free will, and the causality of disease;

TABLE 3–3.  Quality of Evidence of Epidemiologic Effectiveness

| Grade | Weight | Unit of Epidemiological Evidence |
|---|---|---|
| IV | 2 | The opinion of a group of respected authorities based on field clinical experience, *or* a Delphi study, *or* a descriptive study, *or* a report of an expert committee. [However, "respected authorities" are not always correct, nor even reliable; they may be respected, but not authoritative.] |
| IV | 2 | Evidence obtained from a comparison between outcomes at times or places having or lacking the intervention. A dramatic result in an uncontrolled experiment could also be considered. |
| II | 3 | Evidence obtained from a well-designed cohort or case-control analytic study. |
| I | 6 | Evidence obtained from a double-blind Phase III randomized clinical trial. |

*Source:* Adapted from Canadian Task Force on the Periodic Health Examination (1979).

3. conflicts among the needs to satisfy scientific rigor, technical performance, and political and economic realities;
4. the relatively brief time commitment of foreign investigators, leaving residents to bear the long-term consequences, for good or harm, of the trials or procedures;
5. the potential enforcement of perceived beneficence on someone else who may or may not ask for it, need it, or want it;
6. potential competition with, and diversion of resources from, other priorities in health or other areas.

Persons with varied political viewpoints have argued for decades about the existence of a "right to health."

Is health a "right?" If so, is there a corresponding right to medical care? A claim of right ordinarily implies that something may be demanded of another party, such as a government. Obviously each citizen cannot demand from society unlimited efforts to reverse the effects of advancing years or to undo every bodily defect. There must be some baseline that reflects the community standard of a tolerable life. Even if this could be specified, controversy would still flare over whose responsibility it is to achieve and maintain the minimum standard (Basch 1990b).

The Council for International Organizations of Medical Sciences (CIOMS), based in Geneva, has sponsored a number of conferences dealing with ethical issues in biomedicine. Their publication "Ethics and Epidemiology: International Guidelines" (Bankowski et al. 1991) includes sections on drug and vaccine trials and other epidemiologic investigations in developing countries.

## The Greatest Good Argument

*Prevention versus treatment.*

The available information about risks, both those of the disease and those of the preventive and therapeutic means of combating it, seldom allows precise estimates of those risks. If the estimates are very imprecise, it may not even be clear which is riskier, the disease or the preventive measure. But even a clearly lower risk for prevention does not mean that prevention is preferable, since more people are subject to it than to the disease and since the people who suffer from the preventive measure may not be the ones who would have suffered from the disease. . . . The true benefit of any preventive action is not the objective improvements in length of life and in health, but the value individuals place on those improvements, and how they value them in comparison with alternative outcomes (Russell 1986).

*Individual and community interests.* Sometimes there is a conflict between individual and public priorities in establishing policies for health interventions. Assume that an immunization can protect against a serious disease, but at the cost of frequent minor and occasional severe adverse effects. Does the familiar dictum *primum non nocere* (first do no harm) apply to these usually minor injuries inflicted by health authorities? Can the damage be absolved because a greater, though rarer, evil is being protected against? Can individuals be expected, or even compelled, to accept a small risk in the name of a larger benefit for a collectivity such as the community or mankind in general?

In a sense immunization functions like insurance, in which many people pool relatively small payments to compensate those few people who suffer disastrous losses. Premiums are charged in proportion to the anticipated risk. By the same analogy the risk that each individual should assume in prevention can be somewhat higher if the disease is invariably serious or fatal.

Compulsory childhood immunization against smallpox was abandoned in the United States in the 1960s when it became clear that diminished transmission of the virus had shifted the risk-benefit equation so that the risk of harm from the vaccination was greater than the risk of harm from the disease.

Where immunization is not compulsory, it was suggested previously that individuals might encourage immunization in everyone else to minimize their risk of infection, but abstain personally to avoid the risk of harm from the procedure. If everyone were to follow this strategy, then clearly no immunization program could proceed. This conflict in motivations has been considered by Fine and Clarkson (1986), who defined a "total infection-related risk" to cover all risks both of the natural infection and of the vaccine promoted to control it:

A paradox immediately becomes apparent in attempting to define an "optimal" or "critical" vaccine uptake. Optimal for whom? As far as a *community* is concerned, the optimal uptake is that level which minimizes the total morbidity load in the population. But for an *individual,* the concern is more with a critical level of vaccination coverage in the community which minimizes his or her personal risk of disease—below which he or she should elect to be vaccinated, and above which he or she should elect not to be vaccinated. . . . A community's optimal vaccination policy need not coincide with that of an individual, even assuming each to be fully informed and rational. . . . A community's motives in vaccinating an individual are twofold (to protect the individual recipient and to protect his or her neighbors), whereas an individual's motive may concern only his or her protection (Fine and Clarkson 1986).

*Health as a sustainable state: The Maurice King argument.* According to King (1990a) a "demographic trap" exists in which death rates fall and birth rates are sustained, or even rise, so that

> if the birth rate does not fall, the death rate will ultimately rise again, so the population is stuck in the trap and finds itself in an unsustainable state with a high birth and death rate, with ever increasing pressure on its resources, and with a rapidly deteriorating environment. . . . The possible outcomes are limited: the population can (a) die from starvation and disease; (b) flee as ecological refugees; (c) be destroyed by war or genocide; or (d) be supported by food and resources from elsewhere, first as emergency relief and then perhaps indefinitely.

Many environmentalists have written in similar terms for decades about the "population bomb" and related resource-related issues, but King turns his attention specifically to concepts of health and to public health strategies. He argues for measures intended to reinforce sustainability, defined as "the maintenance of the capacity of the ecosystem to support life in quantity and variety."

> The demographic and ecological implications of public health measures must be understood at all levels, especially by the community. If these are desustaining (sustainability reducing), complementary ecologically sustaining measures, especially family planning and ecological support, must be introduced with them (King 1990a).

A new strategy, "Health in a Sustainable Ecosystem for the Year 2000," is suggested by King to replace the existing WHO strategy of "Health for All by the Year 2000" (see chapter 2). Concern would expand from the health of people to the health of the planet as a whole. King goes so far as to suggest that "if no adequately sustaining complementary measures are possible, such desustaining measures as oral rehydration should not be introduced on a public health scale, since they increase the man-years of human misery, ultimately from starvation." This idea recalls the many predictions of inevitable starvation that have been made since the time of Malthus. It flies in the face of contemporary international political correctness and is considered heretical by international health and relief agencies.

## Ethical Aspects of Working in Developing Countries

Well-educated people in wealthy countries are expected to understand the principles of scientific investigation and the need to conduct objective clinical trials. Reiser (1988) remarked that "it is the public's perception that medicine seeks to subject its theories and practices to unbiased evaluation, which separates fact from belief, that is at the heart of its support of the enterprise." Conversely, in parts of the world where scientific principles are low in the public consciousness, broad understanding and acceptance of the rationale behind biomedical field trials may be lacking.

The conduct of clinical trials is made far more complex when extended across national borders. Developed-country scientists working with Third World populations are often accused of "cultural imperialism" or similar transgressions, especially when local people are consulted only marginally or not at all about the program. The logic of field testing a vaccine or drug in localities where the pertinent disease occurs is incontrovertible. Equally convincing are the red flags that go up when a universally applicable innovation is tested exclusively or primarily on developing-country populations. Allegations of unethical conduct are made against those who give the appearance of selecting populations that are docile, uninformed, and unlikely to object to being used as test objects.

> When populations are difficult to obtain in a particular country, either because of the attitudes of the population or the restraints placed upon experimentation, investigators may look to other countries where, for reasons of poverty, ignorance, avarice, or lack of legislative restraints, they can find populations available for experimentation (Robbins 1977).

In any field project, local people may be suspicious of the "real" motives behind the work. Cultural and political viewpoints may combine to arouse hostility in some residents of the host nation. For example, a carefully negotiated agreement was signed between the United States and India in July 1987 to permit the testing of genetically engineered vaccines against rotavirus, cholera, *Shigella, E. coli, Salmonella,* pertussis, typhoid fever, hepatitis B, and rabies. The agreement aimed to protect the interests of all parties because:

> India has always been sensitive on the issue of its people being used as guinea-pigs for the trial of drugs and vaccines developed elsewhere. . . . This difficulty has now been met by the United States, which has associated Indian scientists with the U.S. laboratories where the vaccines are being developed. . . . Allaying fears that the agreement with the United States will make Indians guinea-pigs, [the Indian representative] said the rights and welfare of human subjects of research will be protected, taking into account the laws and regulations in both countries. Clearance would be obtained from the drug controller of India before tests of the vaccines, he said (Jayaraman 1987a).

Despite the official agreement, the proposed vaccine trials encountered a storm of protests. One local scientist warned against trials with imported vaccines before "we are in a position to design and do every trial and test ourselves." Others were disturbed by giving foreigners access to data on epidemiology and immunity profiles of the population (Jayaraman 1987b). An editorial in the *Times of India* called for cancellation of the program. Critics cited the danger of release of genetically engineered organisms, and others were uneasy about patents, copyrights, and other intellectual property. Eventually a group of eight scientists from each country worked out an acceptable compromise (Jayaraman 1988).

A further example of ethical-political turmoil is the debate about the injectable steroid contraceptive Depo-Provera, which was approved for use in the United States in 1992 after decades of controversy.

> It has become a symbol of a wider set of discussions, involving not merely contraceptives, but the intent and execution of family planning programmes generally, as well as the behaviour of pharmaceutical multinational organisations. It has been acclaimed as a much wanted contraceptive, the use of which will protect women from the risk of childbirth and abortion. . . . Conversely, it has been damned as an example of Malthusian enthusiasts foisting unsolicited and questionable therapies on other people hence creating unwarranted risks especially for the poor and those least able to understand the benefit/risk considerations or to defend themselves against commercial exploitation (Potts and Paxman 1984).

*An imputed legacy of unethical practices.* From time to time, some multinational pharmaceutical companies are accused of abuses in developing countries. Accusations include deceptive practices in the marketing of infant formula; exaggerated claims of applicability and efficacy of drugs; insufficient warnings about adverse effects and failure to provide information on packaging inserts; overzealous promotion of products, including bribery of officials and lavish gifts to prescribing physicians; and distribution of drugs that are obsolete or severely restricted in industrialized countries.

The actual occurrence of these alleged ethical infractions is difficult to document. Anecdotal reports may be exaggerated for various reasons. Nevertheless, the imputation of malevolent intent may compromise otherwise valid field trials and the eventual availability of legitimate and useful products.

Does a biomedical innovator with a potentially beneficial product have an ethical obligation to pursue its field evaluation, even if antagonism to the test is expressed for obviously political reasons or because of commercial rivalry? Aside from pragmatic reasons of feasibility and logistics, can there be an ethical justification for *not* testing, or even for abandoning an innovation?

*Ethical aspects of diagnostic products and procedures.* Ethical issues can arise in the case of noninvasive diagnostic procedures, some of which may not even require the presence of the patient. Certain test results may lead to situations in which respect for individual confidentiality and the right to privacy may conflict with legal requirements or a perceived moral obligation to protect the health or safety of others. Individuals may have their activities restricted following a positive diagnosis. For example, certain forms of epilepsy may preclude issuance of a driver's license; typhoid carriers may be prevented from taking up food service occupations; and in certain countries a diagnosis of leprosy or HIV positivity can lead to involuntary physical restriction.

What if we did not know? If a good test is available can we intentionally not use it in order not to incur an obligation to do something we dislike?

Is there an ethical responsibility to inform an individual of the results of a diagnostic procedure, even when no known medical treatment is effective? In some countries patients are routinely not told that they have cancer, which is considered a stigmatizing disease. In fact, most patients are probably aware of the situation and maintain the requisite cultural charade. The case of HIV is different. Although little can be done for the individual found infected, the public interest is served by informing that person of the diagnosis and urging him or her to modify certain behaviors to prevent further infection of others.

On another level, an investigator who has a promising new diagnostic test may wish to conduct a field trial in an endemic area, to compare the innovation against the current standard procedure. Assume that a costly but effective treatment exists. If the trial is carried out in a remote village where no medical care would have been available anyway, what obligation does the investigator have to those found positive? Must they be treated? If funds are available only to test the diagnostic method, but not to treat the patients so identified, should the study not be carried out at all, or should the investigator merely inform those individuals found positive and suggest that they seek appropriate care at their own expense?

## Ethical Issues within Clinical Trials

*Informed consent.* This phrase entered the literature in 1957 in a California Court of Appeals decision in the case of *Salgo vs. Stanford University*. Mr. Martin Salgo had suffered permanent paralysis as a result of a misadventure during a diagnostic procedure to visualize his aorta by injection of contrast fluid. He claimed negligence in the procedure and also cited the failure of the physicians to inform him that the procedure carried a risk of paralysis.

> The court . . . concluded that Salgo's doctors had a duty to disclose to him "any facts which are necessary to form the basis of an intelligent consent by the patient to the proposed treatment." Justice Bray went on to indicate that this *new* duty to disclose risks and alternatives to treatment was a logical extension of the already established obligation to reveal a treatment's nature and its consequences (Silverman 1989).

Standards for obtaining informed consent have varied over time and place. We do not know what, if anything, Edward Jenner said to James Phipps (age 8) or to his parents before the first trial of cowpox vaccine 1796, nor Louis Pasteur's comments to Joseph Meister (age 9) or his parents when he tried his antirabies serum for the first time in 1885. We do have access to the consent form used by Walter Reed in his studies of the transmission of yellow fever in Cuba in 1890, which shows various differences from that in use today.

In the United States the basic document is found in 45 CFR (that is, the Code of Federal Regulations, Title 45), Part 46. This act (Public Law 93-348) also established a National Commission for the Protection of Human Subjects of Biomedical and Behavioral Research. The report of this commission, known as the Belmont Report of April, 1979, states the basic ethical principles and guidelines for research involving human subjects. In sum, these are:

1. *Respect for persons.* Individuals should be treated as autonomous agents. Persons with diminished capacity are entitled to protection.

## CONSENT FORM EMPLOYED BY WALTER REED, 1900

The undersigned, _____, being more than twenty-five years of age, native of Cerceda, in the province of Cerima, the son of _____ here states by these presents, being in the enjoyment and exercise of his own very free will, that he consents to submit himself to experiments for the purpose of determining the methods of transmission of yellow fever, made upon his person by the Commission appointed for this purpose by the Secretary of War of the United States, and that he gives his consent to undergo the said experiments for the reasons and under the conditions below stated.

The undersigned understands perfectly well that in case of the development of yellow fever in him, that he endangers his life to a certain extent but it being entirely impossible for him to avoid the infection during his stay in the island, he prefers to take the chance of contracting it intentionally in the belief that he will receive from the said Commission the greatest care and the most skillful medical service.

It it understood that at the completion of these experiments, within two months from this date, the undersigned will receive the sum of $100 in American gold and that in case of his contracting yellow fever at any time during his residence in this camp, he will receive in addition to that sum a further sum of $100 in American gold, upon his recovery, and that in case of his death because of this disease, the Commission will transmit the said sum (two hundred American dollars) to the person whom the undersigned shall designate at his convenience.

The undersigned binds himself not to leave the bounds of this camp during the period of the experiment and will forfeit all right to the benefits named in this contract if he breaks this agreement.

And to bind himself he signs this paper in duplicate, in the Experimental Camp, near Quemados, Cuba, on the 26th day of November nineteen hundred.


On the part of the Commission:                    The contracting party,

Walter Reed
Maj. and Surj., U.S.A.

*Source:* Quoted by Englehardt, 1988.


2. *Beneficence.* Persons are treated in an ethical manner not only by protecting their decisions and protecting them from harm, but also by making efforts to secure their well-being.
3. *Justice.* Who ought to receive the benefits of research and bear its burdens? This is a question of justice, in the sense of "fairness in distribution" or "what is deserved."

Applications of these three principles were given as (1) informed consent; (2) assessment of risks and benefits; and (3) selection of subjects.

Subpart D of part 46, which deals with children, emphasizes the restriction of experiments with children that involve more than minimal risk (sections 405 to 408).

Although the Belmont Report dealt only with research conducted or sponsored by the then Department of Health, Education and Welfare (later Department of Health and Human Services), it was eventually recommended that the guidelines in 45 CFR 46 be followed by all U.S. federal agencies. Subsequent consideration led to the recommendation that the guidelines form the basis of a model federal policy with a uniform core of regulations applicable to all departments of the U.S. government. This model policy covers more than 13 pages of fine print in the Federal Register (Office of Science and Technology Policy 1986). Current policy of the U.S. Food and Drug Administration can be found in the Code of Federal Regulations, Title 21, Chapter 1, Part 50 (21 CFR 50). Research supported by the U.S. Department of Defense is covered under 32 CFR Chapter 1, Part 219, which discusses certain unique military-related situations.

*Elements of informed consent.* In the years just after World War II, the Nuremberg War Crimes trials demonstrated the need to define standards for judging physicians and scientists who had conducted biomedical research on prisoners in concentration camps. The resultant Nuremberg Code of 1947 stated that human experimentation is justified only under two conditions: when the results benefit society, and when the research is carried out to satisfy certain moral, ethical, and legal concepts. These concepts are (Sass 1988):

1. voluntary consent of the subject;
2. adequate experimental designs protective of the subjects and based on prior animal experimentation;
3. highest technical and professional standards regarding equipment, facilities and staff skills;
4. cessation of the experiment at the request of the subject or experimenter;
5. risk-benefit analysis should accept no risk not exceeded by the humanitarian importance of the problem.

The Nuremberg Code of 1947 established the requirement for voluntary consent of all human subjects. Following this prototype, other basic documentation on experimentation with human subjects includes the two Helsinki Declarations of the World Medical Association. Helsinki I (1964) is a general guide for physicians doing clinical research. With Helsinki II (1975) the emphasis shifted from assessing the performance of investigators to protecting the rights of subjects. Other codes of research ethics promulgated by various national and professional groups are listed in the exhaustive paper of Sass (1988).

## Informed Consent for Clinical Trials

Involvement of human subjects in drug or vaccine trials is governed by multiple and overlapping regulations. Investigators should obtain the specific human subjects guidelines, if any, and request clearance, if required, from all institutions and agencies with responsibility relevant to the trial, including:

- The home institutions of the principal investigator and major collaborators
- The U.S. Food and Drug Administration (or the equivalent national-level agency in participating countries)
- The U.S. National Institutes of Health, U.S. Agency for International Development, or other sponsoring or funding agency
- The World Health Organization or other intergovernmental agency
- The host government Ministry of Health and other responsible national-level agencies
- The regional, state, provincial, municipal, or local health department or similar agency
- Any participating host-country university, research institute, or laboratory
- Participating municipal or community authorities, including legal, customary, traditional or religious leaders

Many of the regulations are formally codified. Others may reflect the customs, practices and traditions of any party to the project. In any event it is prudent not only to adhere to the technical written requirements, but to take minutes at all meetings, annotate discussions at the local level, and initial or sign agreements reached. Clearances and permissions should be obtained in writing.

Subjects enlisted in a clinical trial, or their parents or guardians, are asked not only to accommodate the requests of the investigators but to assume risks of unknown magnitude. Besides the rules and regulations, universal ethical precepts of responsibility, equity, and courtesy must underlie all interactions with human subjects involved in the trial and with their communities.

Regulations concerning informed consent are specified for the United States in CFR Title 45 (Public Welfare), Part 46 (Protection of Human Subjects). Copies are available from the Office of Protection from Research Risks (OPRR) of the National Institutes of Health. The essential principle of informed consent as defined in section 46.116 is that each subject must be provided with the following information, subject to certain exceptions:

1. A statement that the study involves research, an explanation of the purposes of the research, and the expected duration of the subject's participation, a description of the procedures to be followed, and identification of any procedures which are experimental
2. A description of any reasonably forseeable risks or discomforts to the subject
3. A description of any benefits to the subject or to others which may reasonably be expected from the research
4. A disclosure of appropriate alternative procedures or courses of treatment, if any, that might be advantageous to the subject
5. A statement concerning the extent, if any, to which confidentiality of records identifying the subject will be maintained
6. For research involving more than minimal risk, an explanation as to whether any compensation and medical treatments are available and, if so, what they consist of, or where further information may be obtained
7. A person or agency to contact for answers to pertinent questions about the research and research subjects' rights, and a person or agency to contact in the event of a research-related injury to the subject

8. A statement that participation is voluntary, refusal to participate will involve no penalty or loss of benefits to which the subject is otherwise entitled, and the subject may discontinue participation at any time without penalty or loss of benefits to which the subject is otherwise entitled.

Normally the subject's informed consent is documented on a written consent form signed by the subject or the subject's representative, who retains a copy of the document.

*Institutional review boards (IRBs).* The National Research Act of 1974 mandated that each entity applying for a grant or contract for any program involving human subjects must establish an IRB to review all relevant applications. Elaborate requirements were established for membership, procedures, and recordkeeping by these boards. Criteria were specified for IRB approval of research proposals, as described in 45 CFR 46. Institutional Review Boards now exist in health service ministries and departments, and in research institutions throughout the world.

## International Guidelines

International guidelines for biomedical research involving human subjects were proposed a decade ago in a joint report by the Council of International Organizations of Medical Sciences (CIOMS) and the World Health Organization. When research is conducted by investigators of one country on subjects of another, "The ethical standards applied should be no less exacting than they would be for research carried out within the initiating country" (CIOMS 1982). International guidelines for ethical review of epidemiological studies were described in detail by the CIOMS (Bankowski et al. 1991).

Foreign research is discussed also in the 1986 document of the U.S. Office of Science and Technology Policy. In essence, the Model Policy respects foreign laws or regulations that may otherwise be applicable and that provide additional protection for human subjects. Heads of U.S. agencies have the discretion to accept procedures that offer at least equivalent protection for subjects of research carried out in foreign countries. Where several different codes of informed consent may be applicable, investigators as a rule should adhere to the most stringent.

Documents must be presented to potential subjects in an appropriate way, both linguistically and culturally. Consent forms should be pretested for comprehension and conceptual acceptability and should be back translated to English (or their original language) by an individual uninvolved in the original translation to be certain that ideas were properly expressed.

*Informed consent in developing countries.* The informed consent document designed for use in the United States contains idealized language that may be neither appropriate nor practical in many developing country settings. Special conditions apply in the case of persons who for reasons of age, education, or world experience may not be in a position to evaluate potential benefits and risks of experimental procedure. Nonliterate subjects require special efforts to inform them about the purpose of the study, the need for their consent, and their freedom to withdraw. Where signatures cannot be obtained, statements can be read to the subjects, or to their parents or guardians, who may question any aspect of the study. Thumbprints, witnessed written marks ("X"), or oral consent may be substituted for signatures. The situation is often more complex, particularly for persons living in very traditional communities.

Where the concepts underlying experimental evaluation are inconsistent with local tradition, informed consent in the Western model may be simply unobtainable and essentially meaningless whether or not a signature or witnessed mark is obtained.

The significance of cultural differences has been emphasized by Christakis (1988):

> The Western principle of informed consent is predicated on the notions of respect for persons as individuals and as autonomous agents. This is at variance with more relational definitions of the person found in other societies, especially in Africa, which stress the embeddedness of the individual within society and define a person by his or her relations to others. . . . Where the notion of persons as individuals is not dominant, the consent process may shift from the individual to the family or to the community. . . . In some cultural settings it may be extremely difficult to convey an accurate understanding of the idea of randomization or other essential scientific concepts. Moreover, there may be cultural variations in the understanding of disease, at odds with Western scientific notions, that make truly *informed* consent impossible.

Dr. Ebun Ekunwe of Nigeria (Ekunwe 1984) has introduced the term *reverse ethics* for the situation in which too much explanation and trying too hard can keep patients away. He has used the term *uninformed consent* for situations where the germ theory of disease causation is still not accepted. "People are afraid to sign or thumbprint any document. They feel, rightly, that the writer of the document has a hold, usually sinister, on them once they have signed" (Ekunwe 1984).

Similarly, in India it has been stated that

> mere signatures would not ensure the requirements of informed consent. In many instances such a process serves only a ritual function, leaving the patient no more informed or autonomous than he or she would have been if no information had been disclosed. . . . In some communities the very concept of experimental evaluation of therapy is alien and inconsistent with cultural precepts. . . . [The] "doctor knows best" attitude is commonly prevalent in developing countries (Adityanjee 1986).

The Proposed International Guidelines for Biomedical Research Involving Human Subjects (CIOMS 1982) says, "Where individual members of a community do not have the awareness of the implications of participation in an experiment to give adequately informed consent directly to the investigators, it is desirable that the decision whether or not to participate should be elicited through the intermediary of a trusted community leader." This clause has been criticized on the grounds that social pressure may be exerted on any individual who differs from the opinion of the selected community leader, and on the grounds that the leader ("interme-diary") could be susceptible to inducement, bribe, or fraud. The concept of proxy consent at best remains far from the ideal in developing countries (Adityanjee 1986).

*Compensation for participating in a clinical trial.* Walter Reed had no diffi-culty with paying compensation to participants in his trials, but the issue of pay-ment presents many dilemmas in developing countries today. Any substantial ma-terial reward to subjects could easily influence individuals to participate primarily for that reason, negating all pretense of informed consent. The poorest individuals might be the most influenced, bringing a systematic bias into the study and ex-posing the investigator to a charge of exploitation. Similarly, a community leader or intermediary should not be presented with more than a symbolic token for his or her part in the project.

The issue of compensation for participation has at least two aspects: (1) remu-neration to individuals for taking part in the study; and (2) compensation for dam-ages caused, or alleged to have been caused, by some experimental procedure.

Most experimenters consider the payment of compensation to subjects as inev-itably corrupting. Some think that a modest *quid pro quo* should be provided to poor people who participate in a study. Adityanjee (1986) has suggested principles along the following lines:

1. If monetary compensation is to be given, it should not be mentioned until informed consent has been freely obtained.
2. The money should never be of such a magnitude as to be an inducement to join the project.
3. Costs of transportation, food, and loss of wages can be reimbursed if appropriate, and some compensation given for the biological samples taken.
4. There is no waiver of rights to later compensation for damages.

A reasonable alternative to compensation of individuals is to present something to the community for the benefit of all members, both subjects and nonpartici-pants. Athletic equipment, or items for the local school, religious center, or meet-ing hall might be presented as a thank-you gesture at the conclusion of the project, provided that the gift was not a precondition for participation.

In one community in which a diarrheal disease study was conducted, the project directors established a community clinic. Families enrolled in the study, who had to provide frequent stool samples, complained that their nonparticipant neighbors also had access to the clinic. The project manager then started a raffle with prizes of household items such as blankets and pots, giving a ticket in exchange for each stool sample.

*Providing health care benefits to the community.* The 1991 CIOMS report points out that the undertaking of an epidemiological project in a developing country may create the expectation in the community that it will be provided with health care, at least while the research workers are present. It is their opinion that where people need health care, arrangements should be made to have them treated or they should be referred to a local health service that can provide the needed care. The field team should consider carefully what can be done within project budget and protocol to make such services available in a manner that does not jeopardize the scientific integrity of the study.

It is important also to plan from the start to train local health workers in skills needed to improve health services. In this way, "when a study team departs it leaves something of value, such as the ability to monitor disease or mortality rates" (Bankowski et al. 1991).

*Community-based research without individual informed consent.* Research may be undertaken on a community basis. For experimental treatment of water supplies, health services research, insecticide trials, or nutritional supplementation of everyday foods, individual consent may not be feasible. The decision to undertake the research must be made by the public health authorities.

> Nevertheless, all possible means should be used to inform the community concerned of the aims of the research, the advantages expected from it, and any possible hazards or inconveniences. If feasible, dissenting individuals should have the option of withholding their participation. Whatever the circumstances, the ethical considerations and safeguards applied to research on individuals must be translated, in every possible respect, into the community context (CIOMS 1982).

## Research Abroad on Drugs and Biologicals Not Yet Approved in the United States

*Protection of research subjects.* United States Food and Drug Administration regulations dealing with research performed outside the United States are described in 21 CFR Chap. 1, Subpart C. Well-qualified investigators with adequate facilities are permitted to follow the standards of the Declaration of Helsinki or of the host country, whichever is more protective. In 21 CFR 312.110, when a request is made for investigational use in a foreign country of an unapproved new drug or unlicensed biological product, the FDA Commissioner can authorize shipment if he receives, through the International Affairs Staff of the FDA, a request either from the person seeking to export the product or from the government of the country to which it is to be shipped. The formal request from the foreign government must meet certain conditions that are described in the next chapter.

*Drugs for tropical diseases under PL 99-660.* In 1986 the U.S. Congress passed the Drug Export Amendments Act of 1986. This act removes the previous restriction permitting the export of only those drugs (including biologicals) that had been approved for use in the United States. The law is based on the belief that countries have the right to assess their own public health needs and to make their own decisions regarding the products that should be made available to them in order to meet those needs. Section 104 of Title 1 refers to tropical disease research:

SEC. 104. TROPICAL DISEASES. Section 301 of the Public Health Service Act (42 U.S.C. 241) is amended by adding at the end the following: (c) The Secretary may conduct biomedical research, directly or through grants or contracts, for the identification, control, treatment, and prevention of diseases (including tropical diseases) which do not occur to a significant extent in the United States.

According to an unpublished FDA document (Young et al. 1986), the class of drugs that may be exported under this section is limited to those intended for diseases and conditions that are (1) health problems in countries with tropical climates and (2) do not exist to a significant extent in the United States. Under this definition, "tropical diseases would include, for example, parasitic infestations such as intestinal nematodes, trypanosomiasis, leishmaniasis, schistosomiasis, malaria, amebiasis, and filariasis, as well as some microbial infections such as cholera and leprosy."

To export such drugs, an applicant must convince the FDA that the drug is safe and effective for its intended use. Ordinarily there must be data from at least two well-controlled clinical investigations that support effectiveness, as well as data on safety. Therefore, adequate field trials must clearly *precede* permission to export such products.

## The Obligation for Post-Trial Monitoring

Many vaccines or drugs can cause serious damage, or even death, in sensitive individuals. This occurs so rarely that a phase III clinical trial of reasonable size and cost cannot be expected to accumulate enough cases to delineate the risk. Post-release surveillance must be relied upon for that purpose. The product should be withdrawn if shown to be unsafe—but how many cases of adverse effects are needed? How can the unknowable number of deaths prevented be weighed against the known number of severe reactions caused, or presumably caused, by the product?

As an ethical issue, can a field trial of a vaccine or drug be undertaken without an assurance that continuing surveillance will accompany any subsequent routine use of that vaccine? If funds for such monitoring are not in sight when the trial is contemplated, should it be carried out anyway? Is there any obligation to carry out routine monitoring of newly approved vaccines or drugs, and if so, whose obligation is it? The innovator? The manufacturer? The agency that funded the trials? The local public health authorities?

## Fairness to Producers of the Innovation

The attention of ethicists has always been concentrated on the subjects, patients, and other recipients of biomedical services, and little space has been devoted to the issue of fairness to those who develop, produce, test, and distribute products intended to improve the public health. This relative neglect has come with a cost, at least in the United States. The risk of high monetary judgments against companies for alleged harm to recipients reduced the number of licensed vaccine manufacturers from 37 in 1966 to 18 in 1980. By 1989 only three major domestic market vaccine suppliers remained:

In certain instances the U.S. courts have granted damages greater than the total annual production value of products. The highest award so far was for over $30 million to the parents of a 4-year old girl. . . . It seems questionable that any society should on the one hand be committed to public health policies which encourage, or even make obligatory, various forms of vaccination, but have on the other legal provisions which make vaccine suppliers "strictly liable" for all vaccine side effects, even when no negligence is involved and risk warnings have been given. . . . It can strongly be argued that in such circumstances the community as a whole, through its political and public health institutions, should make fair and appropriate provisions for those accidentally harmed during the course of programs designed to benefit the entire population. The fact that in countries like the U.S. this is not the case is one more reason why vaccine manufacture has become increasingly seen as an unattractive area for pharmaceutical industry investment (Taylor and Laing 1989).

This is exactly what has happened, as excise taxes mandated by the National Childhood Vaccine Injury Act caused the public clinic price of the vaccines for routine childhood use to increase greatly.

# 4

# Confirming Vaccine Efficacy in Endemic Areas

It is important to distinguish innovations likely to be useful from those that will probably be unsuccessful. This chapter will describe ways to evaluate vaccines, with emphasis on the conventional standards of safety and efficacy. A third standard, cost-effectiveness, is growing quickly in importance. All governments are more and more concerned with ways in which a stated goal can be achieved more economically. The cost-effectiveness of health interventions is particularly crucial in developing countries.

Relevant working definitions are borrowed from the Office of Technology Assessment (OTA) of the U.S. Congress:

- *Efficacy* is the probability of benefit to individuals in a defined population from a medical technology applied for a given medical problem under ideal conditions of use.
- *Risk* is a measure of the probability of an adverse or untoward outcome and the severity of the resultant harm to health of individuals in a defined population associated with use of a medical technology applied for a given medical problem under specified conditions of use.
- *Safety* is a judgment of the acceptability of risk in a specified situation (OTA 1978).

Moses and Brown (1984) tell us further that

"safety" is obviously a desirable property of any medical technology, but the concept is somewhat subtle. Analysis is likely to show that we always mean by that term 'safe enough, considering.' An effective treatment for a grave disease may entail a considerable risk of mortality, but be accepted with regard to safety because that risk is small compared to the risks of the target disease, and no safer, effective therapy is known. For a headache remedy, the appropriate safety standard is a tiny risk of any dangerous side effects . . . 'zero risk' is an imaginary standard.

# Evaluation of a New Vaccine Intended for Use in Developing Countries

Figure 1–1 shows, in diagrammatic fashion, stages in the genesis of a vaccine for use in developing countries. Formal evaluations, with defined purposes and protocols, occur at several points. The early phases are part of development in the laboratory and extend to preliminary trials in animals. The later phases involve human subjects in clinical and field trials. The entire process requires years to complete.

## Basic Science Phase: Underlying Investigations

*Objective:* Establish feasibility of immunoprophylaxis for the disease in question.

*Strategy:*

   a. Laboratory investigation: Identify potential protective antigens, conduct preliminary protection tests in laboratory animals, identify genes and clone if appropriate, purify and characterize antigens, and perform related procedures.
   b. Basic needs assessment: Through library research and communication with workers in the field carry out an epidemiological and economic analysis of transmission and control in defined situations. Establish the need for the vaccine and the feasibility of immunoprophylaxis for populations in endemic areas.

## Postbasic Science Phase: Definition, Specification, and Organization

*Objective:* Prepare the candidate vaccine for preliminary evaluation.

*Strategy:*

   a. Laboratory investigation: Characterize the antigen physicochemically, functionally, and immunologically in animal models.
   b. Develop descriptive and normative specifications for the product.
   c. Decide on product formulation, including vehicle, carrier if used, adjuvant, preservative, other excipients; consider combinations with other vaccines or immunogens.
   d. Develop procedures for manufacture and analytical methods for quality control.

## Preclinical Trials in Animal Models for Safety, Tolerability, Immunogenicity, and Efficacy

*Objective:* Establish that the antigen and the vaccine cause no harm in laboratory animal models and that they induce acceptable humoral and/or cellular protective immune reaction(s).

*Strategy:*

   a. Select animal models: Proceed from rodents to primates. Select primate species based on pathologic and immunologic similarity to humans as far as practicable.

   b. Safety and tolerability: Study pharmacokinetics, metabolism, and toxicity; look for pyrogenicity, carcinogenicity, mutagenicity, teratogenicity, effects on fertility, hypersensitivity, autoimmunity, granuloma at injection site, and other signs of intolerance; perform routine hematology, biochemistry, EKG (if possible), necropsy, and histology on test animals.

   c. Immunogenicity and efficacy: Monitor humoral and cellular indicators, perform double-blind animal challenge experiments.

## Phase I: Clinical Trials for Safety, Tolerability, Immunogenicity in Human Non-Immune Volunteers in Nonendemic Areas

*Objectives:*

   1. Assess human tolerance and immune responses to vaccines found to be safe, immunogenic, and protective in animal trials.

   2. Determine optimal vaccine configuration and dosage schedules for later phase II and phase III trials.

*Strategy:* Conduct a double-blind, placebo-controlled randomized clinical trial, which may be extended to uninfected volunteers from different ethnic groups. Carry out in the country in which previous work has been done.

   a. Volunteers: Healthy men aged 20–30. Define exclusion criteria; for example: history of the disease or of related infection; residence or travel in endemic areas or other significant probable exposure; evidence of other illness; taking immunosuppressive medication; evidence of HIV infection; detection of preexisting crossreactive antibody titer.

   b. Human subjects protection: Comply with all requirements of 45 CFR 46, including informed consent of subjects and approval by relevant institutional review boards.

   c. Baseline studies: Physical examination; laboratory workup including complete blood count; serum chemistry: liver enzymes, antibodies crossreactive with the vaccine, antibodies to HIV; routine urinalysis; provocative tetanus toxoid to determine immunoreactivity; and other appropriate studies.

*Vaccine:* Prepare vaccines at several antigen concentrations, as sterile single-dose ampules in aqueous saline with or without additives. If lyophilized, prepare diluents. Confirm sterility, nonpyrogenicity, and potency of each formulation.

*Administration:* Take blood and urine for full examination just before the first dose. Divide serum into aliquots for immediate use and long-term storage in ultra-low-temperature freezer. Administer vaccine to several volunteers for each dose according to predetermined schedule.

*Followup:*

   a. Clinical: Observe for immediate toxic effects for 20 minutes. Examine at 24 and 48 hours after administration for fever, local tenderness, erythema, induration, and lymphadenopathy. Ask about headache, fever, chills, malaise, local pain, nausea, vomiting, joint pain, and other symptoms. Two days after each dose do blood count and serum chemistries.

b. Immunologic: Blood samples (randomly coded): at baseline, 1 week after the first dose, then biweekly for 16 weeks. Draw two types: (1) without anticoagulant for serum antibody titers; (2) in heparinized tubes, for cell-mediated immunity studies.
1. Humoral: All tests in triplicate.
   A. ELISA: Determine immunoglobulin groups (IgG, IgM, IgA, IgE) and subgroups against all major vaccine antigens.
   B. Activity against the relevant pathogen in vitro: Set up positive and negative controls with known uninfected and presumed immune sera. Assay effect on organisms by motility, proliferation, thymidine incorporation, or dye exclusion.
2. Cellular: All tests in triplicate. Separate lymphocytes from heparinized blood. Incubate with appropriate antigens with positive and negative controls. Add tritiated thymidine and measure uptake of radioactivity.

## Phase II: Efficacy and Acceptability in Volunteers

*Volunteers:* Select subjects who are (a) human nonimmune volunteers exposed to experimental challenge or (b) human volunteers exposed to natural infection in endemic areas.

*Objectives:* Assess protective immunity and continue monitoring safety and tolerability.

*Strategy:* Expose to (a) known or (b) unknown challenge with the pathogenic organism. Use a blinded randomized clinical trial format, with placebo or other control as appropriate.

*Considerations:* The challenge infection must be carefully monitored, must not cause permanent harm to the volunteer, and *must be curable by chemotherapy.*

*Planning for forthcoming phase III field trials in the preselected site:*

a. Determine test strategy and needs for biostatistical validity at desired confidence limits. Identify computer programs to be used for record-keeping and data analysis.
b. Establish lines of authority, responsibility and liability.
c. Estimate resource needs: materials, personnel, vehicles, supplies and equipment. Prepare budgets.
d. Obtain secure funding for all subsequent phases.
e. Prepare preliminary training materials for field staff and translate if necessary.
f. Prepare drafts of informed consent documents and determine needed clearances for human subjects that will be satisfactory to funding agencies and to international, U.S. and local IRBs, authorities, and regulatory agencies.

## Phase III: Population-Based Trials in an Endemic Area

*Objectives:*

a. Assess protective efficacy of the vaccine in an endemic community setting.
b. Monitor safety and tolerability among vaccinees under noncontrolled conditions in the community.

c. Determine impact of the vaccine on transmission of the pathogen.

d. Adjust vaccine to optimal formulation and dosage schedule.

*Strategy:*

a. Comply with all ethical guidelines for informed consent and human subjects clearances.

b. Recruit predetermined numbers of subjects, usually children, from target communities; utilize valid sampling procedures; conduct preimmunization clinical and environmental studies as appropriate; randomize into vaccinee and placebo groups. Immunize and follow closely for evidence of adverse effects, infection, and/or disease caused by the target pathogen.

c. Study cost-benefit of immunization compared to other means of protection (e.g., vector control) and to case-finding and chemotherapy.

d. Consider integration of the vaccine into primary health care and ongoing immunization programs.

## Phase IV: Post-Registration Surveillance

*Objectives:* Continue monitoring safety and tolerability. Monitor acceptability, usage, and effects on the community. Monitor vaccine impact on the epidemiological situation to optimize strategies for deployment.

*Strategy:* Establish a system for surveillance and reporting of the incidence of the target disease in both vaccinated and unimmunized individuals. Look for late and rare adverse effects of the vaccine.

## Evaluating Diagnostic Technologies

Diagnostic methods can be considered under two headings: those utilized during the evaluation of a vaccine, and those intended as routine methods in the community. Evaluation of a diagnostic tool is relatively straightforward, provided that a "gold standard" exists for comparison.

The primary concerns for diagnostics are accuracy and cost. An inappropriate, misleading, or erroneous test is no bargain even if it is free. Because laboratory-based diagnostic methods demand costly facilities, equipment, and staff training, economies of scale would argue for larger laboratories with lower unit costs. However, that is contrary to the intention of providing diagnostic services for primary health care at the local level.

Quality control is paramount. Tests must be designed to perform well under conditions of prolonged storage and usage under suboptimal conditions by minimally trained persons.

In many developing countries the major existing diagnostic tool at the village level is embodied in the local health worker. Where certain endemic diseases, such as malaria, schistosomiasis, and childhood gastroenteritis are everyday occurrences, they can usually be recognized and successfully treated by minimally trained workers with marginal, or perhaps nonexistent, laboratory facilities. Therefore innovations for diagnosis of such commonalities must be well justified in terms of cost-effectiveness. Foolproof "dipstick" tests designed for primary health care posts could make a significant contribution to ongoing Phase IV disease surveillance.

Although diagnostic needs are just as great in rural areas as in the cities, equipment is generally in shorter supply in the countryside.

> The rational distribution of funds for laboratory equipment and personnel between the large urban centers and rural areas is frequently jeopardized in the budgetary process in less developed countries by political imperatives that tend to shift most of the resources to the urban centers where most of the higher skilled physicians practice and where unmet needs and wants of the population can rapidly produce political consequences. Within health budgets in less developed countries as in developed countries, diagnostics must compete with therapeutic costs which are often extremely high since most drugs must be imported and more importantly with personnel services costs which in many countries often reaches seventy percent of a total health budget (Imperato, 1985).

## Strategic Issues in Use of Innovative Diagnostics in Developing Countries

Presuming that an acceptable diagnostic method is already in use, its replacement by an innovation must be justified on grounds of cost-effectiveness, reliability, accuracy, simplicity, acceptability, and sustainability. Nelson (1986) mentioned that 60,000 people in India make a living by visual diagnosis of malaria on microscope slides. The introduction of DNA hybridization probes or other automated means of malaria diagnosis would displace these workers and would presumably lead to widespread unemployment among microscopists. If the cost per subject and the degree of accuracy are the same, what are the arguments for and against adoption of a capital-intensive technologic innovation to replace a labor-intensive one? For example, is it important to decentralize techniques such as malaria diagnosis to the local facility level? What of potential environmental hazards introduced with diagnostic innovations such as radiolabelled DNA hybridization probes or methods that use toxic materials? These may be justified, but risks should be identified, evaluated, and minimized before the technology is transferred.

## The Clinical Trial

An outline of general considerations regarding vaccine trials, as discussed throughout this volume, is presented in Table 4–1. A great deal of practical information may be found in the handbook by Smith and Morrow (1991), which must be on the bookshelf of every investigator interested in field trials of interventions against tropical diseases.

TABLE 4–1.   Some Considerations for a Vaccine Trial in a Developing Country

A. Regarding the target organism

- Occurrence of different species, geographic strains, local and temporal (seasonal or secular) variants
- Occurrence of multiple stages of the life cycle
- Location within the human host and accessibility to vaccine damage
- Nature of the antigen(s) presented to the host
- Intrinsic reproductive rate and inherent genetic variability
- Relative likelihood of genetic selection for vaccine resistance

B. Regarding the vaccine

- Safety: lack of adverse effects
- Efficacy against essential elements of the target organism(s)
- Nonantigenic components including adjuvant and carrier
- Potency, uniformity, quality control
- Absence of unwanted components, particularly genomic segments
- Shelf life, stability, storage, cold chain requirements
- Manufacture and reliable supply

C. Regarding the participants

- Age
- Sex
- Degree of resistance to organism or restriction of immune capacity due to

    a. genetic or ethnic factors
    b. prior exposure or acquired immunity
    c. current infection with this or other pathogen(s)
    d. nutritional and health status including pregnancy

- Type, number, and timing of other immunizations
- Access to prophylactic drugs
- Measurement of humoral and cellular response
- Comparison with age-matched groups in developed country
- Monitoring for side effects or adverse reactions
- Degree and duration of protection from infection or from pathologic effects of disease
- Establishment of cooperative relationship with mothers

D. Regarding the immunizations

- Route, mode, dosage, and number of inoculations
- Administration separately or with other vaccines
- Total cost of administered vaccine, and cost benefit of its use

E. Regarding the trial

- Source and adquacy of funding
- Site selection: number and distribution of trials
- Approval of international, home country, host country, and local regulatory agencies, as applicable
- Randomized selection of vaccinee and control groups
- Characteristics of co-administered alternative vaccine or placebo
- Ethical aspects

    a. establishment of Institutional Review Boards
    b. informed consent by participants
    c. subsequent administration of potent vaccine to placebo group

- Training field teams and standardizing procedures
- Identifying populations at risk but not already infected or immune continued

TABLE 4–1. (*continued*)

---

- Determining exposure to and risk of infection by target organism
- Statistical validation of data; ensuring statistically required number of participants
- Presence of related or competing pathogens
- Duration and nature of follow-up for

  a. adverse effects (safety)
  b. efficacy

F. Regarding the host country conditions

- Acceptability of the procedure by the residents
- Coordination with and integration into other public health programs
- Effect of vector control or other simultaneous projects
- Determining cost-effectiveness and effective demand
- Competition with other priorities and health programs
- Determining how many clinical trials will the authorities authorize (or tolerate)?
- Establishing relations with donor agencies if needed

---

## Getting Permission to Conduct the Investigation

After all the negotiations have been completed with colleagues and authorities in the host country, investigators in the United States still may need to obtain permission from the U.S. government to export the product and conduct the field trial.

The export of biotechnology products from the United States is governed by restrictions that mandate compliance with the Federal Food, Drug, and Cosmetic Act administered by the FDA, and with the Export Administration Act administered by the Department of Commerce. The DOC can restrict exports for reasons of U.S. foreign policy, short supply in the United States, or national security considerations (Gibbs, 1987). Regulations are different for products that have FDA approval for sale in the United States and those that do not. They differ also for unapproved drugs offered for sale abroad and investigational drugs and biologics intended solely for clinical trials.

*Export of unapproved drugs for clinical trials.* The FDA or equivalent agencies in other countries should be consulted for current regulations (Gibbs 1987). The Code of Federal Regulations [21 CFR 312.110] states that the Commissioner of the FDA may authorize shipment to certain foreign countries of an unapproved new drug or an unlicensed biological product limited to clinical investigation, if a request is received either from the person seeking to export the product or from the government of the country to which the product is to be shipped. If the request is from the proposed exporter, it must include:

- Acknowledgment that the product will be used only for investigational purposes
- Adequate information to satisfy FDA concerns that the product is appropriate for the proposed investigational use in humans [sometimes called a "mini-IND" (Investigational New Drug) application]

- Confirmation from the health authority of the importing country that the product may be legally used by the consignee in that country for that investigation, or a similar assurance
- The quantity of the product to be shipped and the expected frequency of shipments

A request from the government desiring to import the product must include:

- A formal request from an authorized official of the government of the country to which the product is to be shipped
- A statement that the foreign government has adequate information about the product and the proposed investigational use
- Acknowledgment that the product will be used only for investigational purposes
- Acknowledgment that the foreign government is satisfied that the product may be legally used by the consignee in that country for that investigation
- The quantity of the product to be shipped and the expected frequency of shipments

The regulation requires less information from a foreign government than from a commercial firm. The FDA conducts a more stringent review of firm-submitted applications to make a determination that the product is appropriate for the proposed investigational use in humans.

If the FDA authorizes exportation of the product for the purpose of the clinical trial, it will "concurrently notify the government of the importing country of such authorization" [21 CFR 312.110 (b) (2) (i)].

All regulations and legislation may change, so care must be taken to assure that current requirements are observed.

## Comparability of Data

For maximal usefulness, data should be comparable among various projects with similar goals. Where different formulations of vaccines against the same disease are to be field tested, data capture, processing, and analysis should be standardized. Such uniformity would increase the validity of later meta-analysis of the data, but is not often achieved. Agencies such as the World Health Organization are striving to regularize the protocols of different groups to achieve comparability of results.

*Methodologic aspects of interview surveys.* A great deal of health information is obtained by interviewing. A recent review of such surveys concluded that

> the results of health interview surveys (e.g., morbidity rates) appear to be highly sensitive to even fairly minor changes in methodology. When the surveys conducted in LDCs [less developed countries] are reviewed, the most striking observation is the wide variation in methodologies that have been used, for instance, different definitions of illness or morbidity, different interviewing techniques, proxy versus self-responding, and recall periods ranging from the day of interview to 12 months preceding interview. In the absence of any proven and standardized methodology, the interpretation of survey results obtained to date is very difficult, and this has limited their usefulness considerably (Ross and Vaughan, 1986).

In planning data collection by interview survey Ross and Vaughan (1986) recommend that special attention should be given to the following points:

1. Reference population: adequate representation by demographic, geographic, ethnic, socioeconomic, occupational, and other groups as appropriate
2. Study design: sample size, probability sampling, clustering, stratification, estimates of likely errors, characteristics of responders and nonresponders
3. Procedures: background, training, and supervision of interviewers; detection and elimination of bias; pilot testing; checking and feedback of completed questionaires
4. Respondents: whether for self or as proxy for another; who may answer for infants and children
5. Recall period: relation between time interval and importance of events; awareness of and correction for over- and underreporting; memory error; bias; seasonal effects
6. Questions: broad or specific; prompting or open-ended; opinion or fact; comprehensibility; cultural sensitivity
7. Validation: internal consistency checks, reinterview, corroboration by record check or physical examination, repetition with another sample

*Adaptation of programs to local conditions.* In Cameroon, urban areas tend to be reservoirs of measles transmission, and annual measles epidemics begin at the end of the dry season. The measles incidence reaches a peak during the middle of the rainy season and then subsides. Population movements are substantial during this period. Of 6- to 23-month-old children, 27% spent at least one night out of the city, and 4% of the city's children arrived there for the first time during the period of high incidence. The incidence of measles declines at the beginning of the growing season, when many children accompany their mothers to outlying agricultural areas. The pool of urban susceptibles is therefore reduced, but the possibility of transmitting the virus to rural areas increases. The children that escape the disease by leaving return to the city in the fall, thus maintaining a susceptible pool of children for the next epidemic (Guyer and McBean, 1981). No rational measles control program could be planned without knowledge of these elements.

Although overall coverage rates may be high, epidemics can still occur within nonimmunized subpopulations. Attempts should be made to identify pockets of low coverage, particularly in heterogeneous populations. Outbreaks of both poliomyelitis and measles have been reported in well-immunized populations (see chapter 5).

Because of varying difficulty of access, children residing near established health centers are more likely to be immunized. Widely scattered populations oblige vaccination teams to undertake rigorous travel to reach the people. Vehicle breakdown and consequent loss of the cold chain may lead to destruction of vaccines or even to abandonment of the entire program.

## Ending Programs

From the beginning of a temporary program such as a phase III field trial, attention must be given to the time when it will be terminated. In addition to reassignment of personnel, disposition of assets, and financial settlements, program data must be collected and secured and reports written before people become dispersed.

## Dealing with Professionals from Other Cultures

For biotechnologists with little experience in other cultures, the following points may be useful:

- The outsider who respects the local language, culture, and cuisine will generally be well received.
- Westerners should be cautious about adopting or imitating local dress.
- Social interactions are often more stylized in developing countries than at home, and local associates, especially those at a junior level, may be uncomfortable with excessive informality. In some countries it is customary not to include wives at social functions.
- Westerners are often embarrassed at the extent of hospitality shown by colleagues in developing countries. The visitor may be taken to an elaborate restaurant that could cost the host a week's wages. Accept with grace and reciprocate appropriately.
- In some countries a nine o'clock meeting may begin at ten fifteen; in others, your host may be at your hotel room a half hour early, while you are still in the shower.
- Field trials with local populations are sensitive issues in many countries.

In one country where charges of scientific exploitation were made, a collaborative biomedical research program was delayed for months and resumed functioning only after the official bilateral agreement was altered to include the following statements.

- No vaccine developed in the United States can be tested in the host country unless it has already been tested and approved by the U.S. Food and Drug Administration and also by the host country government.
- Any necessary epidemiological research in the host country will be performed only by its own scientists and institutions using their own procedures. U.S. scientists will not be associated with any epidemiological studies in the host country.
- No biological samples collected in the host country can be sent to collaborating scientists in the United States without specific permission from host country authorities.
- Only host country laws relating to intellectual property, copyrights, and patents will be followed.

## Working in Communities

The Primary Health Care norm of community participation assumes the empowerment of a group of people with common interests and comparable goals in health. As Reidy and Kitching (1986) stated in their cleverly titled paper, "Primary health care: our sacred cow, their white elephant," part of the problem for the outsider is ambiguity in the meaning of the word *community*. On the one hand the word may indicate a "place, location, settlement, or a collectivity of people;" on the other, there is an implication of a group of people with coherent and cohesive values and beliefs, with some sort of collective consensus about common health needs, and with a sense of working together for mutual benefit.

While this is often the case, it is not universal. The residents of villages in developing countries are not necessarily more cohesive or single-minded than are similarly sized groups of people who share some legal jurisdiction in North America or Europe. There may be ruling elites and serious conflict based on ethnic group, social class, or political or religious preference, and prejudice and corruption (by Western standards) may exist even within the most photogenic palm-fringed hamlet.

In many countries people in adjacent villages may speak different languages and follow different customs, which disqualifies randomization at the village level, even if environmental and other physical conditions are otherwise comparable. Within-community randomization may also present problems. Despite stratification, it is not usually the case that all who are wanted for the sample can be easily reached or readily convinced that the biomedical innovation offered to them is either necessary or desirable, or indeed that it is not the product of an ominous conspiracy.

## Integration of Innovations with Ongoing Public Health Programs

Vertical programs arise as new health threats emerge and scientific research provides more refined tools to improve health and control disease. Some of these tools, such as immunization and oral rehydration therapy, can be incorporated directly into ongoing public health programs. Others, such as DNA hybridization probes, may be appropriate for more limited deployment. Certain goal-oriented vertical programs, such as onchocerciasis control in West Africa, the EPI, CDD, ARI, and CCCD (see Abbreviations), family planning, or even PHC itself, are considered of sufficient stature to warrant separate organizational structures. Each is distinguished by an establishment of some kind, usually within the Ministry of Health. For each program, staff must be dedicated to planning; to hiring, training, and supervising field personnel; to financing, budgeting, disbursing, and accounting; to management of supplies, vehicles, and facilities; and to monitoring and evaluation. In this way, bureaucracies develop and inevitably serve their own self-interest in addition to their nominal public health function.

*Organization of field trial teams.* Foreign investigators usually cannot establish an entire administrative unit *de novo* to conduct a field trial, but must align themselves with the existing groups that appear most appropriate. In a developing country with few resources, the conduct of a field trial will necessitate the borrowing of facilities, or of trained personnel, from other programs. Even if done on a part-time basis, there will be some effect on the operations of the lending unit. Although both sides may benefit in the long run, it is reasonable that some reciprocation be given, in the form of training of local staff, or donation of vehicles, computers, or other major equipment. In any event, reports and scientific papers should be co-authored with local collaborators.

When vertical programs are successful, as with smallpox eradication, or unsuccessful, as with malaria eradication, personnel and material resources are left without a mission. Such ''orphan'' units may be retooled for other specific functions, such as providing routine childhood immunizations, and may be well suited for the conduct of field trials or other activities associated with technology transfer and diffusion.

*Use of schools.* Schoolchildren can be ideal agents of change. School attendance in poor countries generally stimulates a great deal of pride, and schoolchildren are often eager to apply their knowledge. Many school-age children provide a substantial amount of care of younger children when the parents are unable to do so. Ofusu-Amaah (1983) cites an unpublished EPI study in the Ivory Coast that showed how the instruction of schoolchildren increased vaccination coverage by 6- and 10-fold in two towns.

*Program reinforcement by national authorities.* A supportive national-level health infrastructure always facilitates trials and coverage of routine immunization programs. Coordination by the central government can avert problems such as delayed delivery or heat-damaged vaccine. Collaboration can also promote the image of the immunization effort and increase acceptibility to the community. In Chile, Borgoño and Corey (1978) studied the national experience with poliomyelitis vaccination. Centrally supported health education about polio and its prevention enhanced receptivity at the local level, and reinforcement with biennial massive campaigns provided successful followup.

## Vaccine Efficacy

Vaccine efficacy is a statistical concept usually estimated by a randomized controlled clinical trial. The attack rate of the target disease is determined in groups of equally susceptible and equally exposed individuals who were given, or not given, the vaccine. The level of vaccine efficacy that is determined depends on many elements that can vary from one trial to another, and unwavering credence in any given fixed value is unwarranted (see Table 4–2). Seminal publications on determination of vaccine efficacy are those of Orenstein et al. (1984, 1985, 1988).

### *Major Issues in Epidemiologic Studies of Vaccine Efficacy*

*Assuring vaccine potency and uniformity.* The vaccine to be used in the field trial must be of full potency and uniform throughout the trial. In practice, final large-volume process development is usually not carried out until the success of the product can be assured. After the parameters for efficacy are defined, the vaccine may be modified so that it produces the same biologic responses as did the vaccine that was used in the efficacy trial so that it can be licensed for general use.

Generally samples withdrawn from the field are tested for potency in the laboratory. This is not always a simple matter because techniques to assess the stability of vaccines are not standardized. Most vaccines lose potency on storage, particularly at high temperatures. Some, such as adsorbed diphtheria and tetanus toxoids used separately or as components of DPT, are stable at elevated temperatures even for long periods of storage but may change their appearance and lose potency when frozen (Galzaka 1989).

*Clarifying the concept of a "case".* The epidemiologic importance of a case definition was mentioned in the previous chapter. For many viral and bacterial diseases such as measles, mumps, pertussis, or plague, a case develops when an infective dose of the appropriate pathogen is given to a susceptible host. But just what is an infective dose?

In animal experiments the infective dose can be determined by counting the number of organisms administered to the host by feeding, inhalation, injection, or whatever route is appropriate. Experiments of this kind can be done in humans only with relatively nonpathogenic organisms, and only if the volunteers are carefully monitored and if treatment is available for those who become infected. With viruses and bacteria, the number of infectious particles (organisms) is usually estimated by some sort of standardized dilution method. The infectious dose is often expressed as the $ID_{50}$, which indicates the number of organisms that will cause infection in 50% of susceptible test animals under defined conditions. As usual, it is important to distinguish between infection and disease. After infection has occurred, it is not usually productive to count the number of viruses or bacteria within the body of the infected person that have resulted from the replication of the original invading organisms.

By contrast, in many parasitic worm diseases the organisms enter the body as eggs or larvae, and mature individually to adults. Because adults usually do not replicate within the host, the concept of an $ID_{50}$ loses its meaning. A "case" with ten worms is different from one with 2,000 worms, and neither is comparable to a "case" of measles or cholera. The extent of pathology in most worm infections is not only dose-dependent but also rate-dependent. The larvae may enter all at once, stimulating an acute and severe reaction; or a few at a time over a long period, inducing a protective response but little pathology. The two patterns of infection result in different levels of infectiousness to others, commonly estimated by egg output.

A case definition that poorly represents the disease in question will lead to a biased estimate of vaccine efficacy and can compromise an expensive trial. In many diseases where pathognomonic signs and symptoms are lacking, an ironclad clinical definition is very difficult to formulate. For example, fever occurs in numerous different conditions. Hepatosplenomegaly may be found in hydatid disease, schistosomiasis, malaria, visceral leishmaniasis, and other diseases that may occur in the same area. Additional serologic or other diagnostic criteria may be necessary for an adequate case definition, necessitating the collection and transport of specimens as well as laboratory work that adds to the cost and complexity of a trial.

A dramatic demonstration of the importance of case definition was provided by Blackwelder et al. (1991), who reanalyzed the outcome of a field trial in Sweden of two acellular pertussis vaccines (Table 4–2). Depending on the stringency that one is willing to accept for the definition of a case of pertussis, the determined efficacy of one vaccine (JNIH-6) ranged from 16% to 82%, and that of the other (JNIH-7) could be placed anywhere between 5% and 100%.

> We calculated vaccine efficacy for clinical case definitions based on some combination of the following criteria: duration of coughing spasms (at least 1, 14, 21,

TABLE 4–2.   Case Definition and Acellular Pertussis Vaccine Efficacy

| Case Definition | *JNIH-6 Vaccine* | | | *JNIH-7 Vaccine* | | |
|---|---|---|---|---|---|---|
| | No. of Cases | Risk (%) | VE (%) | No. of Cases | Risk (%) | VE (%) |
| | All Cases Meeting Clinical Criteria | | | | | |
| 1 | 313 | 23.0 | 16 | 359 | 26.2 | 5 |
| 14 | 176 | 13.0 | 22 | 213 | 15.5 | 7 |
| 21 | 94 | 7.0 | 41 | 118 | 8.7 | 27 |
| 28 | 65 | 4.8 | 41 | 60 | 4.4 | 46 |
| 1W | 64 | 4.7 | 39 | 52 | 3.8 | 51 |
| 14W | 40 | 3.0 | 53 | 33 | 2.5 | 62 |
| 21W | 27 | 2.0 | 60 | 26 | 1.9 | 62 |
| 28W | 22 | 1.6 | 58 | 8 | 0.6 | 85 |
| | Culture-Confirmed Cases Only | | | | | |
| 1 | 14 | 1.1 | 75 | 23 | 1.7 | 60 |
| 14 | 11 | 0.9 | 79 | 15 | 1.2 | 72 |
| 21 | 10 | 0.8 | 81 | 12 | 1.0 | 75 |
| 28 | 10 | 0.8 | 78 | 6 | 0.5 | 86 |
| 1W | 7 | 0.5 | 85 | 5 | 0.4 | 89 |
| 14W | 7 | 0.5 | 85 | 4 | 0.3 | 91 |
| 21W | 7 | 0.5 | 84 | 4 | 0.3 | 90 |
| 28W | 7 | 0.5 | 82 | 0 | 0.0 | 100 |

*Source:* Blackwelder et al. AJDC 145:1285–1289. Copyright 1991 American Medical Association

Case Definition = 1, 14, 21, 28 indicate days of coughing spasms.
        W indicates whoops on at least one day.
Risk (%) = Risk of pertussis with the indicated case definition, estimated by the actuarial method
VE (%) = Vaccine efficacy (point estimate).

28, or 35 days), whoops, and nine or more coughing spasms on at least 1 day.
. . . Efficacy was also calculated when culture confirmation was required as part
of the definition. In addition, we calculated efficacy against 21 or more days of
coughing spasms with whoops after adjustment for nonpertussis disease, i.e., cases
that satisfied the clinical definition but could not be verified as being caused by
pertussis infection. . . . An illness was considered a "nonpertussis case" if the
culture was negative, if serologic data were available but no criterion for pertussis
was met, and if there was no documented exposure to pertussis infection (Blackwelder et al 1991).

Situations are known in which illnesses are milder in immunized than in unimmunized children or those given a placebo, e.g., for measles (Aaby et al. 1986).
Therefore some variation in presentation of subsequent illness may be anticipated
in a double-blind study. It may be necessary to search more diligently for cases
or even to draw up diverse case definitions or ascertainment procedures to capture
those modified events that may not occur under "natural" conditions.

In phase III clinical trials of vaccines, repeated surveillance surveys of the total study cohort (vaccinees plus control subjects) will give the least biased estimates but are difficult and costly to do. Complexities of case definition in modified illness following infection of immunized children have been mentioned, and ascertainment procedures might give different results in immunized and control children. For example, it is reported (Blackwelder et al. 1991) that *Bordetella pertussis* is more difficult to culture from immunized then from unimmunized children, leading to possible distortions in conclusions about efficacy.

The use of serology for case ascertainment may be compromised if it is not possible to differentiate between the antibodies arising from natural infection and those from immunization.

Follow-up for case detection and ascertainment depends on the prolonged collaboration of the subjects (or, usually, their mothers) not only in simple things such as answering questions and having temperature taken, but in repeated submission of stool or urine samples and blood specimens.

*Definition of the objective of the vaccine.* The objectives of the vaccine must be clearly stated. Is the purpose to prevent disease in individuals, to control transmission in the community, or both? Vaccine-induced immunity may be intended to block the uptake or establishment of a pathogen (primary prevention), in which case the attack rate (AR) would reflect evidence that infection had taken place. Alternatively, the vaccine may be intended to prevent replication of the pathogen or overt illness (secondary prevention), in which case the AR is defined clinically with all the caveats described earlier.

## Calculation of Vaccine Efficacy

The papers of Orenstein et al. (1984, 1985, 1988) must be consulted by anyone interested in this subject.

## Randomized Clinical Trials

In the classical 1915 formulation of Greenwood and Yule, the ratio of attack rates of disease between vaccinated and unvaccinated groups represents the percent reduction among those vaccinated; hence vaccine efficacy is defined as

$$VE = \frac{ARU - ARV}{ARU} \times 100$$

where ARU = attack rate, unvaccinated;
       ARV = attack rate, vaccinated; and
        VE = vaccine efficacy. The definition and means of ascertainment of "attack rate" must be specified.

The classical formula for VE has been rewritten by Kim-Farley and Sokhey (1988) in terms that may be more practically useful for the situations faced by immunization program managers.

$$VE = \frac{(PPV)\ (PCU)\ -\ (PPU)\ (PCV)}{(PPV)\ (PCU)}$$

where PCU = proportion of cases unvaccinated:

    PCV = proportion of cases vaccinated with the number of doses being examined for vaccine efficacy;

    PPU = proportion of population unvaccinated; and

    PPV = proportion of population vaccinated with the number of doses being examined for vaccine efficacy.

If only one dose of vaccine is required, as for BCG or measles, or where the efficacy of the last dose of multiple-dose vaccine regimen is desired (and the dropout rate between doses is low), the following simplified formula is proposed by Kim-Farley and Sokhey (1988).

$$VE = \frac{PPV\ -\ PCV}{PPV\ -\ (PPV)\ (PCV)}$$

where PCV = proportion of cases vaccinated and

    PPV = proportion of population vaccinated.

The following additional methods for estimating vaccine efficacy are recommended:

    A. *Screening*. When cases of a disease occur in a partly vaccinated population, Figure 4–1 provides a family of curves by which to estimate vaccine efficacy, assuming that the vaccine status of individuals can be ascertained. For example, a community with 50% measles vaccine coverage (PPV) had an outbreak of measles, in which 9% of cases had been vaccinated (PCV). The intersection of the PCV 9 and PPV 50 lines falls on the 90% vaccine 90% efficacy curve, indicating that a good vaccine had been used. This method is only approximate but does indicate whether a more detailed investigation is warranted.

    B. *Outbreak investigation*. Disease outbreaks in defined communities provide an excellent opportunity to estimate vaccine efficacy retrospectively. Door-to-door surveys can be used to determine both disease prevalence and immunization status.

    C. *Secondary attack rate*. This method looks at attack rates, by immunization status, in family contacts of cases with a presumption of uniform exposure within households. Cluster methods may also be employed to investigate secondary attack rates in urban areas.

## Case-Control Studies

These may be the most appropriate means to determine the efficacy of vaccines against diseases such as schistosomiasis. The protocol does not require equal numbers of vaccinees and controls. Some children before the expected age of exposure can be given the vaccine, and others not. The study format could be vaccine versus tetanus toxoid or some other beneficial procedure in lieu of placebo, vaccine A versus vaccine B, or some similar plan. A first follow-up is conducted at a substantial interval, e.g., a year later, with a search for cases. The immunization status is then determined for these cases and for matched, comparably exposed persons who do not have cases. Advantages of this format include lower costs and

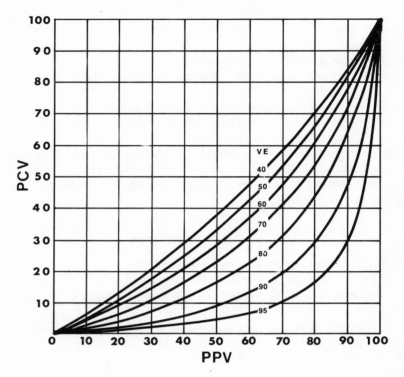

FIGURE 4–1. Percentage of cases vaccinated (PCV) per percentage of population vaccinated (PPV) for seven values of vaccine efficacy (VE). From Orenstein et al. 1984.

avoidance of frequent survey follow-ups, although surveillance must be maintained for adverse reactions. Disadvantages include a fairly long blank period after
immunization, the need to identify matched controls and assure that they do not
have cases, and a means of objective determination of immunization status. Vaccine efficacy in a case-control study is usually calculated by using the odds ratio
(OR) to approximate relative risk (RR) according to the following formula.

|              | *cases* | *controls* |
|--------------|---------|------------|
| *vaccinated*   | a       | b          |
| *unvaccinated* | c       | d          |

$$RR \cong OR = \frac{ad}{bc}$$

$$VE(\%) = (1 - RR) \times 100$$
$$= 1 - \frac{ad}{bc} \times 100$$

A major problem of case-control studies of vaccine efficacy is that when the disease is common in the vaccinated group (if the attack rate among the vaccinated (ARV) is greater than about 20%) the odds ratio fails to be a good estimate of relative risk and the imputed vaccine efficacy may be too high (see Orenstein et al. 1988 for correction and other formulas).

*Generalizability of determined efficacy of a vaccine.* Phase III clinical trials are costly and cannot be repeated many times. There is no guarantee that any one randomized clinical trial will yield *the* true answer. For example, PRP-D conjugate vaccine, administered to infants to prevent disease caused by *Haemophilus influenzae* type b, was found by Eskola et al. (1990) to be "highly effective in protecting young Finnish children . . . against invasive *H. influenzae* type b infections." In another study of the identical vaccine, Ward et al. (1990) concluded that there was "no evidence that the PRP-D vaccine provides significant protection, at least for Alaska native infants, against invasive diseases caused by *H. influenzae* type b." A third study of the identical vaccine (Siber et al. 1990), determined that antibody levels to *H. influenzae* type b were lower in Apache Indian children in the United States after immunization than in white children before immunization. In another example, oral cholera vaccine CVD103 proved to be more immunogenic in U.S. and Swiss volunteers and Thai university students than in poor Thai soldiers and children (Taylor 1991). Contrasting conclusions about the efficacy of a single vaccine after trials employing similar protocols may reflect ethnic and genetic differences in the populations studies, or a variety of other confounding factors.

At least ten large randomized placebo-controlled clinical trials have been conducted to assess the protective effect of BCG against tuberculosis:

> These trials have involved tens of thousands of subjects who have been followed after vaccination for 20 years or more. Most of the trials have been conducted in North America, but one was in Puerto Rico, two in South India and one in the UK. All of the trials have included an unvaccinated or placebo group and usually estimates of vaccine efficacy have been based on substantial numbers of cases of tuberculosis. The trials have given estimates of efficacy ranging between zero and 80%. . . . It seems clear, however, that the variations are unlikely to be due to methodological artifacts and it must be recognised that the efficacy of BCG varies substantially in different parts of the world (Smith 1988).

In partial explanation, BCG is actually not a unitary vaccine but a variety of products made in many countries using genetically variable organisms. Therefore some variation in efficacy may be expected, but the range from 0% to 80% appears extreme.

*Trial results versus routine use.* There is no guarantee that the efficacy determined in any single trial will be identical to the efficacy of the same vaccine used routinely, even within the same population. Many factors may intervene; for example, when management and supervision of the trial cohort is insufficient; mothers may not adhere to the strict immunization schedule; the age of children immunized may vary widely; immunizations may be given only in periodic campaigns; vaccine may not be stored as carefully; injections may not be given as accurately with regard to dosage, location, or technique; different vaccines may be "bundled" or administered together; local genetic strains of the pathogen may change; environmental and nutritional conditions in the communities may differ because of climate, economic condition, hostility, or disaster; and follow-up, case detection, and reporting may not be continued with the diligence of the trial.

## The Monitoring of Adverse Clinical Events

As immunizations are given, a method must be put in place to detect and report adverse reactions. Such monitoring within a clinical trial differs from routine postlicensing surveillance of vaccines and immunization programs, which is discussed in chapter 5.

It is important not to apply post hoc reasoning to all adverse events following immunization; that is, to assume that because some medical incident occurs following an immunization, that the immunization is the cause of the event. Some events are genuinely coincidental rather than consequential, although it may be difficult to convince a family, or perhaps an entire community, that that is the case. Spurious associations, or even ficticious and unfounded rumors, may be difficult to eliminate from the public mind.

Adverse events are often grouped into immediate (24- to 48-hour) and longer-term. Fever, soreness, or irritability (in infants) are relatively common. Mention of this possibility should be made in the informed consent statement in advance of the immunization. Other events are less predictable, but more significant, and should be intentionally sought. When the time interval after the intervention is very prolonged, it is difficult to establish a clear connection, an extreme example being illness from HIV infection transmitted by a contaminated syringe or needle.

Adverse events may result from an unfortunate biological interaction between a particular vaccine preparation and an individual child, or from a technical error such as improper dilution, poor injection technique, or microbial or chemical contamination. The provocation of paralytic polio by properly administered injections has now been well established (Wyatt 1984; Sutter et al. 1992).

The various methods by which adverse events may be monitored, and the advantages and disadvantages of each, have been tabulated by Cocchetto and Nardi (1986). The two general strategies for monitoring are subject reporting and professional follow-up.

## Subject Reporting

*Spontaneous reporting* of adverse events has several advantages. "Leading" the patient with unintentionally suggestive questions is unlikely. The subject is presumably under no pressure to report, literacy is not required, and both predictable and unpredictable events can be detected. On the other hand, individual initiative is necessary to make a report, and the subject must recognize and discriminate the information to be reported.

*Diary card.* In this method the subject maintains an account in his or her own words to describe the events following the intervention. Again, events both expected and unexpected by the investigator may be uncovered. However, literacy is required of the respondent, who may feel obliged to report something. Diaries tend to elicit large amounts of unstructured information, and events may be underreported or overreported. Diary cards can be illegible, misplaced, or lost. Personal contact and conversations with the subject are lacking.

*Self-assessment check-list.* A pre-prepared format requires little patient initiative and allows self-rating of severity as mild/moderate/severe or on some sort of scale. This format requires little staff time to administer. Disadvantages are that "leading" the patient is more likely because only certain items dealing with more likely and predictable events are listed, and literacy and motivation are required.

*Professional follow-up.* These methods may include observation by the investigator, or by a member of the health team such as a visiting nurse or health auxiliary. The focus is on detecting objective events, with the advantages that conversation and rapport with the subjects is facilitated; subject literacy is not required; specific opportunities are presented for respondents to provide data with minimal initiative; and both predictable and unpredictable events can be detected. Disadvantages of direct observation are the staff time and expense required for intensive observation; possibly inadequate detection of subjective events; and differences among various observers (observer bias) requiring rigorous training, standardization of procedures, and checks for comparability. Depending on the phrasing of the questions, the subject will need to decide what must be reported and may feel obliged to report something. On the other hand, the subject may feel intimidated or may have inadequate incentive to volunteer information.

Direct observation with or without physical examination can be combined with formal check-lists and questionnaires. These assure interactions between investigator and subject and tend to result in a higher frequency of intervention-related events than verbal probes. Great care is needed in wording questionnaires so that they do not detect only predictable events, or obscure important adverse effects among trivial complaints.

## Biotechnology, Vaccines, and Environmental Risks

From the earliest days of molecular biology, citizens, government officials, and scientists have expressed concern about the safety of the laboratory techniques and the recombinant organisms produced by them. A group of scientists closely involved in the development of recombinant DNA technology, spearheaded by Paul Berg of Stanford University, directed organized attention to the potential hazards at a conference in Asilomar, California, in 1975. Although the obvious hazards of work with Lassa virus, HIV, or similar dangerous agents necessitate continuing containment, most official assessments of the environmental risks of routine research using recombinant DNA have permitted relaxation of the elaborate requirements mandated during the immediate post-Asilomar era of the mid-1970s. Nevertheless, there is a continuing need to assess the potential for adverse effects caused by the release or escape of organisms or materials into the environment, and of any subsequent disturbance of natural ecosystems. In addition to escapes or releases, there is the possibility that some biomedical innovations, used as intended, may have unexpected adverse environmental consequences. For this reason Recombinant DNA Advisory Committees (RACs) have been established at the National Institutes of Health and at many other institutions and agencies with responsibilities in these areas.

If introduced organisms could be guaranteed to do their job and disappear, much apprehension would be dissipated. In reality it is probable that most would not survive in the long term because of evolutionary mechanisms controlling the balance among species in natural environments. Pests and pathogens, together with diminished capacity for survival in the wild, are likely to prevent most recombinant organisms from becoming established and may block the natural transfer of organisms from one habitat to another.

Nevertheless, the biological literature is replete with examples of organisms intentionally or accidentally introduced into a foreign area with disastrous results: African bees in Brazil, rabbits in Australia, mongooses in Hawaii; starlings in North America; the fungi causing Dutch elm disease and chestnut blight in the eastern United States; Mediterranean fruit flies in California; and influenza viruses around the world.

Biologists, politicians, and the general public in many countries are aware of the potential hazards of importing and releasing living organisms. When these are products of genetic engineering, people's spontaneous reaction is likely to range from caution to hostility to condemnation. Among people who do not even know the words, concerns are expressed about toxicity, carcinogenicity, teratogenicity, and undefined but insidious effects on the environment. On the other hand, expertise in laboratory biomedical science or technology confers no special competence to assess environmental risks. The fact is that nobody knows what, if anything, is likely to happen. A group of eminent scientists was questioned about their views on this subject by Alexander (1985), who remarked:

> I do not think I have ever before seen such a collection of noted scientists, including geneticists, molecular biologists, biochemists, microbiologists, and even a few ecologists, who drew firm conclusions without citing any evidence or providing any documentation. . . . They were saying, in effect, that either there was a catastrophe coming, or, alternatively, that the likelihood of there being a problem was zero.

The degree of uncertainty about these matters is so high that credence should be given to reasonable concerns about escaped or released organisms:

- The possible replacement or even extinction of pre-existing species
- Unplanned dissemination through water courses, by wind, or carriage in or on phoretic hosts

- Inability to retrieve or neutralize the organism in nature
- Further mutation or reversion to increased virulence
- Population explosion owing to lack of natural enemies
- Harmful effects on non-target organisms including humans
- Dispersal of altered genetic material and potential transfer to other organisms
- Financial and resource costs that might be incurred for control

Unreasonable fears and suspicion of malevolent intent may be fostered by science fiction, horror films, and literature, just as two centuries ago cartoonists depicted horns and cattle heads growing on persons inoculated with cowpox virus. As mentioned elsewhere, political motivation may be a powerful force in condemning field trials before they are even started.

The sense of apprehension is not abated even among the most scientifically aware group in highly educated Japan. The self-selected readers of the Japanese science magazine *Newton* responded with suspicion to a recent survey about biotechnology:

> Forty percent mistrusted vegetables bred by tissue culture; almost three-quarters were dubious about the prospect of genetically engineered fish coming onto the market. Sixty percent . . . worried about using engineered bacteria to kill pests. . . . A staggering 90 percent of the Japanese science buffs surveyed rejected researchers' claims of environmental safety, though only 31 percent would go so far as to advocate banning free-release experiments. . . . But most depressing are the answers to *Newton's* last three questions: 57 percent of the science-reading public felt that bioresearchers' safety claims were unreliable, 88 percent averred that "biotechnology researchers and institutions are concealing information which is potentially damaging to their fields of study." And 77 percent predicted that biotechnology will develop into the same sort of "major social problem" as atomic energy—which is, in Japan, saying a lot (McCormick 1989).

In recent years the clamor regarding potential environmental hazards of released organisms has abated as the predicted disasters have failed to materialize. The biomedical enterprise is not nearly so heavily involved in such matters as are agricultural, mining, or marine biotechnology sectors. Nevertheless, those concerned primarily with health must be sensitive to the possibility that environmental considerations might result from their efforts.

## Defining the Environment

To non-Western people, it may make little sense to discuss "the environment" as a neatly distinct package separated from other aspects of life. To hunter-gatherers or agriculturalists, what we consider an environmental risk may be viewed as not particularly different from any other kind of hazard within the continuum of the physical, spiritual, personal, family, and community dimensions of life. Even for scientific observers, the interconnectedness of environmental and other issues is apparent. For example, some have compared the potential ·of new biotechnology with the old biotechnology of the "Green Revolution."

The economics of Green Revolution agriculture tended toward large scale mechanized monocrop production which, while feasible for big land owners, was seldom possible for small land owners who were often, in effect, forced off their land. . . . There were ecological costs in the contamination of waterways by chemical run-off from heavily managed fields. In Southeast Asia, for example, small fish and crustaceans in the waterways and rice paddies provided an important protein supplement in the diets of rural people. Green Revolution chemicals destroyed this source of food, and the grain that they produced was characteristically shipped to the cities and even for export. National trade balances were improved and interest on international debts was met with less difficulty, but the disadvantages of the old biotechnology seriously diminished its net benefits (Caldwell 1988).

## Environmental Consequences of Living Viral Vaccines

Experience with live Sabin oral polio virus has shown a small but quantifiable frequency of reversions to virulence, with some reports of adverse effects in individuals who became secondarily infected from revertants of live polio vaccine virus originally administered to someone else. For example, the Associated Press ran a story in many newspapers on Jan 25, 1992 about a 48-year-old man who contracted polio from a revertant virus acquired while changing his niece's diaper. "He was staggeringly unlucky to get it from the vaccine, unlucky to be so severely affected and unlucky to have missed vaccination when he was little," said Dr. Marin Wale, a specialist on communicable disease at the hospital.

In the case of living engineered viruses or bacteria, environmental effects conceivably could occur if the organism, used as an oral vaccine, is spread via feces onto the soil and into natural waters. It is also possible that a movable genetic element could be transmitted from the engineered organism to a natural microbial inhabitant of the host intestine or of the environment. Such dissemination could conceivably lead to infections in invertebrates or nonhuman vertebrates. Much depends on the propensity to reversion and the ability of the organism to survive in the environment.

## Vector Control and Pest Insect Management

Of proven value, vector control is a key strategy in reducing transmission of many diseases including dengue and other arthropod-borne viruses, malaria, African and American trypanosomiasis, and filariasis. In addition to their role as vectors, insects are responsible for substantial reductions in crop harvests, direct adverse effects on domesticated animals, and enormous losses of stored foods of all kinds. Many similar techniques may be used for control of insect vectors and insect pests.

Integrated pest management (IPM) is an effort to reduce pest damage to acceptable levels through use of natural predators, pathogens, and parasites; changes in habitat or environment; and limited use of chemical pesticides. The genetic manipulation of insecticidal bacteria such as *Bacillus thuringiensis,* or insecticidal products based on recombinant production of its toxin, could become important elements of IPM programs to control human and animal (or even plant) diseases. Released into the environment, these materials could have undesirable effects. As a result, public opinion could be turned against entirely unrelated elements of disease-control projects, affecting the testing or implementation of environmentally innocent innovations.

## Potential Environmental Effects of Research and Manufacture

*The laboratory environment.* Research groups that study pathogens will need to have stocks or cultures of these organisms, as well as infected animals for vaccine testing and challenge, on hand in the laboratory. Some pathogens, such as *Coccidioides immitis* or *Trypanosoma cruzi*, are notorious for causing infections in laboratory workers or even in visitors to the building or compound. Mice or other laboratory animals may escape, with their infections, into the environment. Laboratory waste, including cage bedding and animal carcasses, must be burned or decontaminated and properly disposed of.

Early concerns about the possibility of accidental ingestion or infection with recombinant organisms such as the workhorse *E. coli* K-12 led to several conferences which concluded (1) that there was little or no potential for these organisms to colonize the human intestine, to be transmitted directly or indirectly from one person to another, or to produce any adverse health effects; and (2) even so, certain precautions should become institutionalized in laboratories working with recombinant DNA. These concerns led to the definition of levels of containment. At the low end, experiments involving the transfer of genes from harmless organisms into *E. coli* K-12 require only normal laboratory precautions. At the other extreme, the P4 level was established for manipulations of genes from major pathogens of humans and other dangerous work. This highest level required the use of negative-pressure safety cabinets to prevent the escape of any materials. Specially filtered air entered the facility, and exhaust air was filtered or even flamed. Water drains were guarded and monitored to prevent backflow or contamination. All laboratory waste was autoclaved or otherwise sterilized. Entrance and exit to the laboratories required passage through decontamination areas.

In agriculture, more than 200 releases of genetically engineered plants have occurred without incident. The elaborate safeguards for microorganisms have, in comparison, often been criticized as excessive. The fact is that no serious escapes have occurred from laboratories so equipped, and both the profession and the public have been reassured to the extent that laboratory safety of recombinant DNA work is now much less discussed.

*The external environment.* The unintentional release or accidental escape of recombinant organisms such as undebilitated *E. coli,* or of plasmids or fragments of genetic material, should be avoided, particularly the accidental transfer of genes for toxin production, or of antibiotic-resistance traits. The disposal of large amounts of fermentation broth from recombinant DNA production or even autoclaved culture media could cause problems of enrichment or eutrophication of streams, waste disposal ponds, or other habitats. The result may be algae blooms, fish kills, and other unpleasant and conspicuous effects.

Research, pilot plant, or industrial-scale installations in developing countries must adhere to good manufacturing practices and not be permitted to skimp on containment, sterilization, effluent control, waste treatment, staff training, or other procedures to minimize environmental risk. Erosion of official or public confidence in the facility or its products may follow a serious environmental episode, with disastrous effects on trials in progress or acceptance of finished products.

In the United States the National Environmental Policy Act (NEPA) (Public Law 91-190, 40 CFR 1500-1508) requires environmental impact statements before certain types of work, particularly with infectious organisms or toxins, may be undertaken.

## Ancillary Environmental Effects of Utilization of Innovations

Implementation of an innovation may generate predictable environmental hazards, such as in the disposal of radioactive materials used in DNA hybridization probes, photographic chemicals, or other toxic laboratory supplies. Disposable syringes and needles, vials, packing materials, and other wastes will be generated during vaccine usage. In areas of high HIV and hepatitis B prevalence the use of unsterile syringes and needles must be prevented. One "obvious" solution to the contamination problem is injection devices that can be used only once. Numerous configurations of self-destructing syringes have been proposed, and several types have been manufactured in the West, but they have found no market in developing countries because of their high cost. Many types of plastic "disposable" syringes and needles, which are routinely discarded after one use in wealthy industrialized countries, can survive several rounds of sterilization by boiling or autoclaving. These items are commonly reused in poorer countries until the markings wear off or the plunger begins to leak. An autodestruct syringe, which is several times the cost of a conventional "disposable," can actually multiply the cost per immunization several dozen times if reuse is taken into account. The volume of material that must be received, packed, transported, and discarded is correspondingly higher if single-use syringes are ordered. These factors are highly important where roads are poor and vehicles, fuel, money, and other resources are scarce.

Reused syringes and needles, when properly sterilized, pose no risk of infection to vaccine recipients. The likelihood of improper sterilization is reduced by the issuance of proper equipment and by adequate training and motivation of immunization personnel, but little can be done in the face of outright fraud. Kristof (1993) described a rural factory in China in which more than a million used hypodermic needles were merely rinsed with pump water, air dried, repackaged in plastic bags, and sold as new. Several persons are known to have died from infections as a result. Although these needles were used in blood transfusion rather than immunization, "the episode offers a window into the challenges that a developing country faces on the way to a market economy" (Kristof 1993).

*Unpredictable environmental side effects.* A highly potent and effective family of new drugs, the avermectins, are used to control nematode and arthropod parasites in many species of domestic animals. Much of the drug administered to cattle transits the intestine unabsorbed. The manure from cattle treated with ivermectin is resistant to breakdown by dung beetles that live on the fields, preventing normal recycling of nutrients from the manure and interfering with the ecology of pasture lands (Wall and Strong 1987). Other products may also have unanticipated effects.

## International Code of Conduct on Biotechnology

Many organizations, such as the OECD, have long held policy conferences about the release of genetically engineered organisms. An informal Working Group on Biotechnology Safety, "composed of representatives from the United Nations Industrial Development Organization (UNIDO), United Nations Environment Program (UNEP), the WHO, and the Food and Agriculture Organization (FAO) has been meeting to consider issues related to biotechnology and the environment. During several meetings in 1991 the group prepared a draft code, which attempts to capture the minimum commonly accepted principles in regard to the release of genetically modified organisms into the environment. Obligations of researchers, governments, and other parties are stipulated, without interfering excessively with the progress of biotechnology. According to UNIDO (1992) the guidelines will:

- Promote research and development and environmental applications of genetically modified organisms
- Provide guidance to national authorities to make quick decisions on proposals for introductions
- Help industry to commercialize products based on genetically modified organisms
- Bring transparency and avoid trade barriers
- Facilitate consumer confidence and acceptance

The extent to which these guidelines will apply to products such as genetically modified live vaccinia viruses (see Appendix B) or, indeed, to oral polio vaccine, is not entirely clear.

## Genetic Side Effects of Some Innovations

Overuse of particular antibiotics is known to stimulate selection of resistant microbial strains that, when disseminated in communities, reduce the effectiveness of the antibiotic. It is possible also that a vaccine effective against a particular type or strain of pathogen may reduce the frequency of the target organism and promote dissemination of less susceptible organisms having other undesirable properties. Second-order effects of enhanced frequency of such pathogens, perhaps within reservoir host species, are essentially impossible to foresee.

## "The Biotechnologic Chernobyl"

Environmental pollution and "sharp deterioration of the environmental situation" forced the former USSR to close at least eight large factories that produce much-needed animal fodder as single cell protein (SCP) from petroleum, natural gas, methanol, and agricultural waste. By 1986, microbial protein production had reached 1.6 million tons. Environmental safeguards in these factories were poor. Despite continuing complaints of nearby residents, one plant near Leningrad continued to release a reported two metric tons of protein dust onto the surrounding community each day. These discharges caused thousands of cases of allergy, including permanent invalidity and many deaths. The resultant public outcry led to closure of all the SCP plants (Rimmington 1989).

The biomedical innovator may see little connection between the mammoth SCP factories and the production of diagnostics and vaccines; but to the public mind biotechnology is biotechnology, and widespread protests in the USSR have resulted in the abandonment of plans for factories that produce perfectly innocent amino acids, antibiotics, and biological preparations.

# 5

# Fulfilling the Technological Promise

## From Innovation to Product

### *Science and Technology (S&T) Policies*

The purpose of a policy is to help make informed decisions in order to achieve objectives in a systematic and economical manner. The essential function of such a policy is to define the range of alternatives, their expected social and economic costs, and the relative desirability of each, from the point of view of the policy-maker. The existence of a thought-out policy does not guarantee success, but its absence foretells inconsistency, ad-hoc and poorly justified judgments, and possible conflicts of interest.

The conception of science and technology for development has changed since the 1950s and 1960s. Then, the wealthy countries operated a supermarket of knowledge and techniques, from which the poorer countries could pick and choose useful items off the shelf under Western tutelage. Recognizing these shortcomings, the United Nations convened the Conference on Science and Technology for Development (UNCSTD) in Vienna in 1979. The intention of the conference was to give priority to the development of science and technology capabilities in Third World countries, and to restructure international scientific and technological relations.

Typical of contemporary comments from the developing countries themselves were those of Wionczek (1979), who characterized the system as wasteful because:

1. Scientists and technologists in less-developed countries participate only marginally in higher educational activities.
2. The demand of the productive system for technical know-how and innovation is satisfied mainly from abroad. Whether public or private, national or foreign-owned, productive units prefer foreign-originated technology, which is viewed as less risky.

3. Local financial and managerial resources are largely inadequate.
4. Available funds are misapplied: up to three quarters of R&D expenditures are for salaries and wages, and local scientists and technologists "tend to follow the 'overequipment' trends fashionable in some advanced countries."
5. Although most existing R&D institutes are short of researchers, the number of institutes proliferates for reasons of prestige, and the best brains are absorbed into administrative, political, and bureaucratic activities.
6. S&T development is unbalanced as to sector and discipline because of haphazard and uncoordinated initiatives, and because decisions are made by foreign-trained personnel influenced by the kinds of research in vogue in more advanced countries.
7. In most less-developed countries more S&T staff are interested in "pure" research than in applied research and technological innovation. But funds for basic research are very scarce and those for R&D are insufficient because of limited interest by industry.

Much progress has been made in the intervening years. Brazil, Egypt, India, Kenya, Mexico, Nigeria, Thailand, and South Korea have been particularly active in establishing S&T policies. For most poorer countries the observations of Wionczek remain valid, and the program of action adopted at the Vienna Conference has, in general, not been implemented.

*Health technology policies.* Decisions about adopting innovations are made only in part by governments. Regarding immunization programs in the public sector, many health ministries follow the policies and procedures of the World Health Organization's Expanded Programme on Immunization and of the United Nations Children's Fund (UNICEF), which oversees much of the bulk purchase and distribution of vaccines and related supplies. The views of donor agencies, where these exist, are also taken into account. In the private sector, much control is in the hands of exporters, importers, and distributors of products, physicians and pharmacists, and families and individuals. The ultimate success of the innovation will rest not on its technical merit, which is a necessary but insufficient condition, but on the decisions made by potential purchasers and consumers of the technology. As the focus narrows from the nation to the individual, the range of accessible options becomes progressively smaller, and decisions to use or not to use are made more on an ad hoc basis rather than through any predetermined policy.

Despite the urgency to obtain the optimum health benefit for increasingly limited investments, few developing countries are able to evaluate biomedical innovations. Decisions about the application of biotechnology are particularly difficult to make, because of its relative novelty, technical complexity, undefined cost-benefit-risk ratios, and indeterminable economic, environmental, epidemiologic, demographic, and cultural consequences. In short, those charged with making decisions often do not know enough to do so effectively.

Conventional descriptions of governmental policy making rely on formal econometric models, based on hypothesized need and adequacy, benefit, and cost. However, insiders such as Banta and Andreasen (1990) know that;

> policies are not made for rational reasons or because of economical arguments. Policies are made because of pressure. People want something. They vote, they

lobby, or they make financial contributions. Organizations also try to affect policy-making. These actions produce pressures on legislators and government officials. Policies result from very complex pressures and political processes.

With a similarly pragmatic outlook, Contractor and Sagafi-Nejad (1981) viewed recipient-country government policies on technology transfer:

> Specific instruments have included laws, registries, review boards, and various other forms of control for gate-keeping and facilitating foreign investment and licensing. . . . In many such cases, the government regulators have made quiet exceptions to their own published rules, if the technology is sufficiently desired, making such rules in several LDCs mere points of departure for purposes of negotiation.

Policies and their application reflect the political realities of time and place, fundamental governmental attitudes, and the interests of different parties. Some of these interests are shown in Table 5–1.

*Elements of health technology policy.* From the point of view of a government, any technology should be thoroughly appraised, and adopted only if it:

- Supports national economic development objectives
- Improves domestic technological capability and self-determination
- Increases the value of the country's human resources
- Fosters regional development and decentralization of production of goods and services
- Reduces dependence on other foreign technology

TABLE 5–1. Health Technology Transfer Incentives of Actors in Developing Countries

| Actor | Relation to Technology | Desired Outcomes |
|---|---|---|
| Government | Gatekeeper | *Promote:* Social equity, use of local materials, employment and training, local investment<br>*Restrict:* Capital-intensive and dependency-producing methods, repatriation of profits; negative cost-benefit, environmental damage |
| Local industry | Supplier | *Promote:* Profit, widespread use, reduced dependency on imports<br>*Restrict:* Unfair advantage of large transnational corporations; excessive regulations and control |
| Health Professionals | User | *Promote:* Selection, safety, efficacy, productivity, profit (private sector)<br>*Restrict:* Complexity of use, ambiguous outcome; access by nonprofessionals |
| The public | Recipient | *Promote:* Choice; physical and social well-being, convenience, low cost, quick and positive outcome<br>*Restrict:* Individual and environmental damage, social disruption, cost and effort |

Some assurance should be given that the adoption of any particular technology will help promote social equity, and not strain further the disparities among social classes. Safeguards could be enacted, or at least attempted, so that the new technology is not used principally by those who can best afford it, but need it least.

Above all, a health technology should increase the productivity of the health sector by providing more health per unit of resources invested than the previous way of doing things.

The Advisory Committee for Health Research (1986) of the WHO investigated ways to match health needs of developing countries to new or emerging technologies. That panel specified the requirements for a user country, discussed at length in the original document;

1. The need for political will
2. The importance of the existence of an appropriate infrastructure for the transfer of technology
3. The need for a national health technology policy
4. Training of personnel
5. Equipment-related requirements

A good policy should foresee potential inequities and balance the interests of various parties in order to further societal goals. If the implementation of any technology should generate adverse epidemiologic, economic, environmental, political, or cultural consequences, the policy should state, in broad and general terms, what should be done to rectify the situation. The following broad elements can be incorporated:

1. Priorities: Planning and decision making
   A. Deciding on health needs
   B. Evaluating available resources
   C. Selecting options and ordering elements on the basis of national philosophy, goals, and principles
   D. Building domestic capabilities
   E. Strengthening health services, human resources, financial capacity, regional competence; decentralization
   F. Freedom from bias or abusive influence
2. Feasibility: Limits and constraints
   A. Technical feasibility
   B. Cost-benefit; economic analysis
   C. Resources: Money, facilities, supplies, equipment; qualified medical, scientific, and technical personnel
   D. Cultural acceptability
   E. Agreement with international standards
   F. Sustainability
   G. Political realities and power structure
   H. Legislation, decree, or regulation
   I. Incentives and disincentives: duties, taxes, subsidies
   J. Enforcement
3. Integration and coordination
   A. Temporal: Institutional memory and archive; analysis of past and present policy-related situations; projection to the future

B.  Governmental: Administrative and political, central, provincial
C.  Intrasectoral: Medical faculties, medical societies, health trade unions, private clinics, HMOs, etc.
D.  Intersectoral:
    a.  Public: education, agriculture, finance, planning, commerce, etc.
    b.  Private sector: industry, banking, venture capital, foundations, etc.
    c.  Public and private links
E.  International:
    a.  Official: Intergovernmental organizations
    b.  Private: Transnational corporations, donors, foundations

4. Promotion of innovation
  A.  Stimulating interest in relevant institutions
  B.  Diffusing scientific and technical knowledge
  C.  Awarding health-directed R&D contracts
  D.  Providing funds for investigator-originated research grants
  E.  Providing scholarships and training opportunities

5. Monitoring and Follow-up
  A.  Surveillance
  B.  Evaluation, control, and adjustment

## Regulating the Importation of Technology

Guidelines to regulate the importation of health technology are difficult to design. They must be in harmony with national science and technology policy, national industrial development policy, and national health policy. Among the many questions that must be answered for effective policy formulation are:

- How can national health care needs be identified?
- What is the role of technology in meeting those needs?
- How will technology transfer be regulated and coordinated?
- What will be the roles of the government and private sectors, professional organizations, universities and research institutes, domestic and multinational commercial interests, public interest groups, multilateral and intergovernmental organizations, and others interested in technology transfer and assessment?
- How can *ex ante* appraisals be done before the technology is deployed?
- How should technology assessment be carried out? Will there be predetermined benchmarks or planned "milestone" reviews? How will quality assurance be determined?
- Who will be responsible for training of appropriate manpower?
- Will sufficient funds for research and development be appropriated in national budgets?
- How will strategically important industrial sectors be promoted?
- What is the place of product importation versus local production?
- Who will own the legal rights to the technology?
- Who will make marketing decisions?
- Who will make ethical decisions about testing the technology on local populations?
- Who will assess the cost-effectiveness of the technology after it is imported?
- Who will be responsible for policing the technology in the event that serious adverse effects are discovered?

*Technologic paralysis.* The rapid pace of technologic development may induce caution in adopting any particular technology because of the likelihood that a better or cheaper one may be just around the corner. The institution that has just invested a million dollars in a new diagnostic scanner may be dismayed to see the next generation of scanners advertised three months later at half the price and twice the efficiency. At some point in a free market a decision must be made to implement a particular technology, but the thoroughness of the pre-commitment evaluation must be proportional to the degree of subsequent obligation to the technology.

## Policy-Setting Bodies

In general, the broader the policy, the higher the governmental level responsible for making and implementing it. Strategic plans and comprehensive policies may be made at prime ministerial or cabinet level; 1- or 5-year midterm plans and enabling policies at ministry level; operational plans and implementing policies at department level; and tactical plans and policies for day-to-day work at program level. Vertical integration within ministries, and intersectoral integration among ministries, are needed for coordination and to avoid conflicting policies and regulations.

Through a formal understanding, those units accountable for making and carrying out policies should be specified, but they must also be given the resources and trained personnel to carry out their mandate. It is impossible to maintain control over haphazard policy making, which is the inevitable consequence of urgency, pressure, and the absence of appropriate plans.

A coordinating group (board, council, panel) for health technology policy should include intersectoral representation, probably at subministerial or departmental-head level. Technologies considered exclusively within the health sector (e.g., dealing with malaria control or nutrition) may nevertheless have implications for agriculture, environment, public works, industry, education, or other sectors, and technologies that primarily affect other areas may similarly impinge on health. The Ministry of Planning, or its equivalent, should therefore be involved.

*Functions of technology policy-setting bodies.* The health technology policy body must be chartered with a specified scope of work which may include any combination of the following responsibilities.

1. Set policy only. Define priorities and objectives, write a general statement of lofty principles, and then go home.
2. Conduct case studies of specific technologies on request, (a) within the government, or (b) for industry and private health institutions, to provide information and analyses on which others can base a decision.
3. Advise about the choice of products or processes, either in response to specific requests, or proactively on its own initiative.
4. Act as a coordinating agency and fund the assessment of specific technologies by government and academic research institutions, as (a) contracts to carry out work

determined to be needed, or (b) grants to investigators applying on their own initiative.

5. Carry out postintroduction surveillance and monitoring of selected applications of technology. Receive complaints and report problems to appropriate government agencies for enforcement if needed.

Specific cases or examples may come to the attention of the group by:

1. Petition by a foreign manufacturer, distributor, or exporter applying to the customs department, ministry of industry, etc
2. Request from another branch of government
3. Request from end-user health professionals or institutions
5. Selection by the board or panel itself

## Prospects for Commercial Viability

The unpromising economic situation in many developing countries inhibits most commercial companies from undertaking research and development on products that are unlikely to be profitable. Pharmaceuticals and diagnostics for tropical diseases are typically far less lucrative than products made for the flourishing pill-taking markets of the United State and other industrialized countries. Much research on drugs and vaccines against parasitic diseases, for example, is directed primarily toward the more profitable veterinary market, and products for humans sometimes follow in the wake. The resultant neglect of major human health problems does not indicate any malignant intention, but merely reflects economic reality. No matter how potentially useful the innovation may be in a technical or humanitarian sense, it can have no practical impact unless it becomes an economically viable product. This subject is explored in Table 5–2.

Complex innovations, such as vaccines and diagnostic reagents that must be licensed, require a major financial investment for research, product design, scale-up, pilot production, testing, and eventual bulk manufacture. Substantial funding is necessary even if the innovation will be in the public domain and not patented or owned by a profit-making company; indeed, it will be even more difficult to secure financial backing for a noncommercial product that is not *intended* to be profitable. In such a case, funds for initial production, and for implementation of field trials, may be sought realistically only from donors such as major foundations, bilateral development aid agencies, development banks, and multilateral organizations.

Innovative technology is transferred to developing countries through collaborative institutional agreements with industrial country partners for testing, production, and marketing of products, or with the assistance of donor agencies and organizations as mentioned above.

## Protecting Intellectual Property Rights (IPR)

Often in the excitement to get a venture going, entrepreneurs get deeply involved with the research and development only later to find out that there are encum-

TABLE 5–2.  Considerations for Commercialization of Biomedical Innovations in
Developing Countries

| Issue | Obstacles and Needs |
|-------|---------------------|
| Regulation | Meeting requirements for approval or licensing<br>Patents and associated disputes<br>Customs and importation issues |
| Safety and efficacy | Demonstrating clinical or other benefit<br>Demonstrated lack of immediate and long-term adverse effects<br>Possible unknown or undetectable hazards<br>Concern over mutagenesis, carcinogenesis, teratogenesis<br>Environmental ramifications |
| Production | Reliable procurement of product or components<br>Stability, uniformity, quality control<br>Scaling up volume of production |
| Logistics and delivery | Assuring a storage and distribution system that will protect the physical integrity of the innovation |
| Local marketing | Identifying epidemiologic and marketing targets with resources to purchase<br>Stimulating professional and community familiarity, acceptance, and demand |
| Marketing to donor agencies | Meeting international regulations<br>Volume of production; distribution network |
| Cost and profitability | Forecasting a favorable investment/return ratio |

brances on it. . . . He or she is interested in the pursuit of knowledge. But ignoring the source of the technology and of its development can cause problems later on in attracting subsequent commercial investment, let alone potential lawsuits for patent infringement (Frank 1985).

*Patents.*  There are two main vehicles to protect an innovation. The first is patent protection: an exclusive right granted by a government to an inventor to exclude others from making, using, and selling an invention within its jurisdiction for a period of years (17 years in the United States). In the United States, the power to enact laws protecting the intellectual property of authors and inventors is embodied in Article I, Section 8, Clause 8 of the U.S. Constitution. Specific regulations vary, but as a rule, full and complete disclosure of the invention must be provided in exchange for issuance of the patent.

Patents have been in existence for a long time, but regulations and conventions made decades or even centuries ago are stressed to the limit when challenged with decisions about modern high technologies. In the pharmaceutical industry, patents are of two general types: on the process, which protects the method of manufacture of a product; and on the product itself, which protects the rights to manufacture, distribute, and sell it. However, in many cases it is not even clear whether the innovation constitutes a process or a product; some examples of ambiguous cases are a method of using a product, a diagnostic kit, a genetic material that has been altered, or living organisms or pure cultures of organisms that do not occur in nature.

More than 100 countries are signatories to the Paris Convention of 1883, which established the principle of equal treatment for domestic and foreign inventors and the acceptance of international priority based on the date of first filing of the patent application in any country. Even so, Agarwal estimated in 1978 that 80 to 85% of all patents granted by developing countries were held by foreign companies or individuals. That percentage remains in the same range today.

Although some argue that IPR protection allows the patent owner to overprice products and inhibit competition, on balance it is likely that patents promote innovation by stimulating the entire research base and facilitating development of all technologies.

*Patents in the United States.* Biotechnology patent law has become extremely specialized, as described in detail by Hofer (1991). The U.S. Patent Office alone has thousands of undecided biotechnology applications on its shelves, principally for lack of trained examiners who understand both the technologic and legal issues involved. Many PhDs in molecular biology have been hired by the U.S. Patent Office as examiners to evaluate applications. Others have gone to law school so that they can advise industry about the legal aspects of their products.

Almost all biotechnology patents that are issued are contested by other companies because, aside from the very rare true breakthroughs, all use essentially the same laboratory processes and procedures. It is also evident that biological materials intended for a specific function in humans, for example, vaccines against the same pathogen, must by their nature have certain similarities.

Adjudication must extend not only to conflicting and overlapping patent applications but to subsequent lawsuits claiming infringement of patents already issued. Because of these complexities, delays of at least several years are the rule if proprietary technology is to be transferred. A company may not want to release a product, or even to reveal a potentially useful method, before at least some degree of patent protection has been secured.

*IPR and technology transfer to developing countries.* Patent systems are probably valuable for the larger developing nations that have a significant research sector, but have little real utility elsewhere. Nevertheless, patent systems are becoming more widespread in the developing world. This proliferation is stimulated partly by outside diplomatic pressure, primarily from the United States, and partly from internal national forces.

Many Western basic scientists are only dimly aware of the international aspects of IPR. When asked about intellectual property, they may interpret the concept only in terms of agreements for co-authorship of scientific publications arising from joint research. In fact, IPR issues are the subject of serious and prolonged negotiations among governments, and if not resolved may turn a promising collaborative effort in technology transfer into a legal and political morass. Complications in protecting the intellectual rights inherent in production processes may represent a major impediment to the distribution of an innovative product in some countries. Nowhere is this more true than in the case of health-related products made by biotechnologic means.

Differences in national policies regarding patents for biotechnology inventions focus on three areas:
The standards governing patentable subject matter with respect to inventions of biotechnology;
The characterization requirements for biotechnology inventions; that is, the choice of morphological, physiological, immunological, or molecular descriptions to meet the varied disclosure requirement for patentable subject matter;
The scope of protection afforded a biotechnological invention that is both self-reproducing and contains biological information usable in other biotechnology innovations (Cook 1989)

The major IPR issue centers around differences in governmental philosophies reflected in the diverse legal systems, especially business, trade, and patent law, among the countries involved, and the consequent need to invest great amounts of time, effort, and money. First, the length of patent protection is as little as 5 years in some countries. Second, a country may offer patent protection only to processes, not products; in that case a slight alteration in production conditions might yield a product essentially identical to another whose production method is described openly in the home country patent disclosure. In addition, the prior publication of research data in the scientific literature can have profound, and varying, effects on subsequent patentability in various jurisdictions. Third, laws and regulations vary concerning patentable subject matter. In some countries, certain commodities considered to be for "the general good" of the people are simply not patentable. The intent of governments that have such laws is to make these products available at the lowest cost for the benefit of their citizens. Many countries also prohibit the patenting of living organisms. All of these issues are complicated further by protectionist policies and trade barriers erected in some countries against importation of certain classes of good.

A new Indonesian Patent Act came into effect in August 1991. This act provides a patent term of 14 years (plus one 2-year extension), and specifically declares that "inventions for production of food or drink, new types of varieties of plants and animals or processes for cultivating plants and animals, as well as the products of these processes, and inventions for methods of treatment or diagnosis of human or animal body, are not patentable."

A clear example of an international dispute about differing IPR regulations arose between the United States and Thailand, largely over Article 9 of Thailand's patent law, which established an "exclusion list" consisting of pharmaceuticals, food products, beverages, agricultural machinery, and biotechnology. Threatening reprisal under Section 301 of the Omnibus Trade and Competitive Act of 1988, the United States demanded the removal of pharmaceuticals from the exclusion list and the extension of all patents from 15 to 20 years. Professor Sirilaksana Khoman, a Thai, commented that

We support the patent system. The U.S. negotiators should therefore refrain from their tiresome reprimand and endless innuendos, treating the copying of life-saving drugs as analagous to common thievery. . . . It should also be made clear that the U.S. government's demands and threats are unjustifiable on both legal and moral grounds. . . . Pharmaceuticals, in particular, have consistently been excluded from patent protection in most developed countries because of their importance to human existence. Extension of patent protection to pharmaceuticals has been relatively recent, even in countries whose level of economic development far exceeded Thailand's at the time. . . . There is no standard law and no standard time for patent protection of pharmaceuticals (Khoman 1991).

Pharmaceutical Manufacturers' Associations in industrialized countries, which are often economically and politically powerful, may be angry that some pharmaceutical companies in certain developing countries may, under local laws, produce drugs patented elsewhere. At the same time those companies may manufacture drugs on which patents have expired and export them as generics to the very same industrialized countries.

The biotechnologic innovator who has developed a product intended to improve the public health in developing countries may step unwittingly into such a hornet's nest, or even into a dispute arising from entirely different issues. For example, retaliatory action may be taken by one country against another because of contravention of copyright laws, software piracy (other manifestations of IPR), alleged dumping of products from shoes to steel bars, or sale of primary goods to an unfavored country (e.g., jute to Cuba). Trade representatives may hold entirely unrelated products hostage during negotiations, or use science and technology (S&T) agreements as protectionist levers or for political influence.

It is much more difficult to "pirate" a product of biotechnology than a cassette or video tape. A recombinant vaccine cannot be reverse engineered as easily as some chemically based pharmaceuticals have been, and it seems unlikely that many groups would have the resources and motivation to do so. Nevertheless, where local companies, or even governments, are free to duplicate or imitate products unprotected by intellectual property legislation, where sales prices are rigidly controlled, and where the outlook for profits may be marginal at best, foreign companies are understandably reluctant to enter the market. In such a case there may be serious consequences for commercial transfer of other kinds of biotechnology (for example, in the field of agriculture), for unrelated products of multinational companies that also have a hand in generating biomedical innovations, and for retaliatory sanctions by developed-country governments.

*Circumventing IPR regulations.* One potential exception to IPR regulations should be noted for the record. In some countries vaccines are manufactured in government laboratories, rather than by commercial producers. Homma and Knous (1992) have noted that in some cases the transfer of technology for production of DPT and BCG vaccines has been made through public institutions and "has not necessitated special arrangements to protect intellectual property or patent rights."

*Where IPR issues may become significant.* Despite the potential pitfalls and complications just described, it is likely that IPR will never cause difficulties in the least developed countries, where there is little indigenous scientific infrastructure, where manufactured goods are all imported, and where international donor agencies are the sole source of an innovative product. It is in the more advanced developing countries, which have a scientific and technologic infrastructure, capable local industries, and a substantial market, that protection of intellectual property rights, together with S&T policies in general, may become significant.

Some of these issues are covered under the complex series of understandings included in the General Agreement on Trade and Tariffs (GATT). In many cases bilateral agreements exist, and others are always under negotiation, but sometimes pairs of countries simply fail to achieve a consensus.

Recent trends in legislation on this subject are discussed by Barton (1989), who mentions:

> significant change in the area of national technology transfer legislation under which governments of developing nations scrutinize technology import transactions. Although the detailed restrictions vary, typically, the laws limit the size of royalty that can be charged for a license and prohibit certain license provisions. . . . The legislation, enacted in many developing nations during the 1970s, aims both to reduce the costs of importing technology and to protect local technology-based firms from foreign competition. In some cases, however, the legislation has been so strict and has permitted such a low return that few international firms are willing to enter into technology supply transactions. The trend now is for these rules to be applied less strictly and, in some cases, even formally weakened.

*Free trade areas.* The reverse situation to restrictive and even punitive trade restrictions is found in the free trade areas, of which the European Community is the best known example. Regional associations such as the Andean countries have formed free trade areas among themselves in which products move freely and no tariffs are charged. The United States, Mexico, and Canada have such an agreement, with a goal to extend the free trade agreement to the entire Western Hemisphere.

*Trade secrets.* The second means of protection is by trade secret or "know-how," information that derives its economic value from not being generally known, and which is the subject of reasonable efforts to preserve its secrecy. Trade secrets depend on concealment and therefore demand a great effort in maintaining confidentiality. They offer no real protection from disclosure and may be revealed for various reasons at any time. Even though laws against fraud, industrial espionage, and theft of intellectual property may apply, their enforcement across international borders presents major headaches, particularly through the legal systems of developing countries. The protection offered by a trade secret does not extend to independent development of the idea by another person or group, nor to discoveries made by reverse engineering of a product obtained on the open market.

*Other official regulation.* In addition to special IPR and S&T regulations, the laws, statutes, decrees, and guidelines of governmental agencies regulate general commercial transactions in all countries. The basis and justification for regulation of biotechnology-derived products is usually the protection of the public health and welfare, including the environment.

Three basic approaches to regulation were described by Brown and Douglas (1992). The first employs fully trained scientist-regulators, as in the United States. The second, found in the United Kingdom, divides responsibility for review and consent between individuals with scientific expertise and others with administrative responsibility. The third system involves individuals who are primarily administrators but are trained in an academic scientific discipline. They can call on persons within or outside government who have special technical expertise. Such a system is used in France.

Within the United States, the Office of Science and Technology Policy (OSTP) of the White House published a "Coordinated Framework for the Regulation of Biotechnology" in 1986 that listed the federal agencies with responsibilities for approval of the products of commercial biotechnology and specified the regulatory jurisdictions of each. The Environmental Protection Agency (EPA) acts under authority of the Federal Insecticide, Fungicide, and Rodenticide Act (FIFRA) and the Toxic Substances control Act (TOSCA). The U.S. Department of Agriculture (USDA) relies on the Plant Pest Act and other statutes. The Food and Drug Administration (FDA) of the Department of Health and Human Services has broad regulatory power derived from numerous federal laws.

The scientist-regulator charged with management of a vaccine should have an understanding of laboratory and research issues, insight into the clinical need for the vaccine and the problems and benefits associated with its use, and detailed knowledge of the regulations. "It should be obvious by now that this person must have the wisdom of Minerva, the knowledge of Einstein, and the patience of Job" (Brown and Douglas, 1992).

Almost all of the literature dealing specifically with worldwide regulation of biotechnology transfer is concerned with Europe, Japan, and other industrialized countries. The European Community adopted uniform regulations in March 1990 when the Council of Ministers approved two directives. The first calls for notification of authorities of all twelve EC member nations whenever genetically engineered organisms are to be used in a laboratory or manufacturing process, and specifies certain containment and emergency procedures. The second directive requires that member states notify the European Commission whenever a genetically engineered organism is to be introduced into the environment. Most biotechnology companies consider these regulations to be too stringent.

Regulatory initiatives in the least developed countries are not well evolved, but biomedical innovators interested in working on Third World health problems should seek out existing national regulations and guidelines about the following topics, particularly in more advanced countries such as Brazil, Chile, Egypt, India, Mexico, the Philippines, Thailand, or Venezuela: testing and distribution of vaccines, biologicals, and pharmaceuticals; recombinant DNA research; protection of human subjects of research; and environmental protection.

Regulatory inconsistencies abound. For example, a biotechnology research program was established by the U.S. Agency for International Development in a large Middle-Eastern country. The funded laboratories ordered chemicals from abroad, including restriction enzymes needed to carry out recombinant DNA studies. When the shipments began to arrive at the airport, federal health inspectors, applying their official procedures manual, classified the enzymes as vaccines, which necessitated removal of a sample for testing by the National Vaccine and Serum Institute. The enzymes, frozen in dry ice, had been purchased in such minute amounts that no reasonable sample could be taken; and in any event, the Vaccine and Serum Institute had no idea about how to test them. The situation was finally resolved by creating a new category in the procedures manual, after which the restriction enzymes were passed without difficulty.

## Informal Regulation and Surveillance by Public Interest Groups

Consumer and private watchdog groups are common in North America and Europe, but rare elsewhere. Such movements are growing in other countries and may come to play a role in the acceptance and diffusion of biomedical innovations. For example, in 1990 a group of Brazilian academicians, pharmacists, and physicians established a group called the Sociedade Brasileira de Vigilancia de Medicamentos (Sobravime) to encourage ''scientific and technical work to help improve the use of medicines and to provide a forum for thorough discussion of the issues involved'' (Carlini and Herxheimer 1991). Despite the support of national and local medical societies, professional associations, and even government agencies,

> The newly founded society was immediately attacked by the Brazilian Association of the Pharmaceutical Industry (ABIFARMA), which represents most of the U.S. and European companies operating in Brazil. . . . ABIFARMA wrote privately to the Ministry of Health accusing Sobravime of trying to institutionalise activities competing with those of the established authorities and undermining their credibility, which 'could lead the public to panic unpredictably,' and of intending to 'steal confidential documents' submitted by companies to the ministry for regulatory purposes. . . . This attempt to kill an open and much needed new organisation demonstrates the ethical backwardness of the Brazilian drug industry and its fierce attachment to the prevailing pharmaceutical chaos in Brazil (Carlini and Herxheimer 1991).

## Importation versus Local Manufacture of Products

Many people in developing countries have the opinion that any products made locally must be inferior to similar items imported from abroad. This unfortunate tendency is sometimes shared even by government officials, who, for whatever reason, may favor more costly imported goods over domestic ones. In some cases the restrictive purchasing policies of bilateral donor agencies do not permit them to provide locally made goods even to public sector projects that they fund within developing countries. Where standards for quality, uniformity, and consistency are met, there are many good arguments in favor of local participation in product development, production, and distribution to the extent reasonably possible within the country.

Wherever the product is made, scale-up and production problems must be solved so that it becomes continually and reliably available. Pharmaceuticals and biologicals destined for use in developing countries may reach their destination via:

1. Importation of the finished product, packaged for the end user
2. Importation of the finished product in bulk, with packaging at a local plant or facility
3. Importation of a product intermediary or precursor, with a local finishing process involving a chemical change, followed by local packaging
4. Importation of raw materials, with local processing, finishing and packaging
5. Local production from raw material to finished product, but using some imported technology, machinery, equipment, and materials
6. Complete, self-sufficient, independent local production.

For different products, or for any one product at different times, one or another of these patterns may be used, depending on:

a. The expected size of the market, based on the population to be served, the prevalence of the condition, and anticipated preferences of the epidemiological or administrative target individuals
b. The presence of national or local facilities including storage and distribution networks, laboratories, clinics and hospitals
c. The relative cost of each alternative
d. The economic situation and availability of hard currency for importation of foreign products
e. Sudden or unexpected need, as with an outbreak, epidemic, or appearance of a condition unusual for the country
f. Local expertise in processing and production, or the desire to obtain such capability through experience
g. Local environmental laws and regulations mandating containment facilities for pathogenic organisms, disposal of infectious, toxic, or radioactive byproducts or wastes, or the like
h. Customs duties and regulations regarding imports and exports
i. Laws or policies requiring the purchase of specified products by government health facilities
j. Laws and regulations regarding patents, the transfer of technology, and local ownership of business enterprises
k. The role, responsibility, and authority of various governmental and private interests (such as customs agents and importers) regarding the decision to make or buy

Adaptation to the local situation can produce epidemiologically and economically effective products that support the public health goals of the nation. Local manufacture may result in cost savings, provided that the product volume and product quality are high enough, and can have significant secondary effects such as promoting expertise in production and management. For such reasons, and to conserve foreign exchange, government funds may be expended primarily for items produced nationally. The private sector may have fewer restrictions on the source of biomedical supplies. Where external donors underwrite projects such as field trials of drug or vaccine efficacy, they are likely to control the source of the materials. In general, outside donors (except for mission hospitals and the like, on a small scale) do not fund ongoing needs on a permanent basis.

In a discussion of problems facing vaccine producers, a spokesman for a state-owned company in Indonesia remarked that

> Technology transfer in turn requires government commitment, adequate facilities, skilled staff, a strong domestic market, financial resource and a regional role for the recipient firms. The problems such companies were likely to face included the high cost of production facilities, increasing technological sophistication and stringent, time-consuming regulatory requirements. Collaboration would be needed to cope with these problems (Children's Vaccine Initiative 1991).

A representative of a major U.S. producer did not share the opinion that vaccine manufacture could be accomplished in developing countries, believing that

> developed countries would in all probability be the source of future vaccines and these would therefore be geared towards the needs of the developing world. Unless special funds were made available, these vaccines would be beyond the means of developing countries. Industry faces several problems in vaccine development, including poor patent protection, stringent price regulation, a growing market in copies of sometimes doubtful quality, and falling profits. Industry in developing countries could handle the filling stages of vaccine production using local labor and currency, but . . . bulk production would have to remain in the hands of manufacturers elsewhere (Children's Vaccine Initiative 1991).

Similar doubts were raised by Vandermissen (1992), who stated that transfer of technology will not make the supply of vaccine any easier or cheaper:

> The most fundamental rule for economical production of vaccines is to manufacture them on a very large scale. . . . Therefore, to duplicate or multiply vaccine production units by definition goes against this very important principle. The only cost factor that will be lower in a developing country, compared with an industrialized country, is the labour factor. Yet labour is not a very important component of total costs, as modern day vaccine production is not labour intensive. . . . On the other hand, in developing countries, the cost of equipment may be significantly higher. . . . To justify a project of transfer of technology on financial or economic grounds is thus difficult. If there must be a justification, it will be found in the moral obligation of the developed nations to help developing nations towards progress.

*Evaluating the acceptability of a proposed vaccine: the WHO standards.* The WHO Biologicals Unit acts as adviser to UNICEF on matters relating to the quality of vaccines offered for use in immunization programs. The WHO Expert Committee on Biological Standardization considers the procedures for preliminary assessment of acceptability, for consistency of production, and for inspection of vaccines proposed for use in United Nations immunization programs. The conclusion of the Expert Committee are published at approximately annual intervals in the WHO Technical Report Series, in which the specific requirements for each type of vaccine (and other biologicals) are set forth in considerable detail.

## Effects of the Importation of Technology on Local Research

Problems of imported technologies (not commodities) have been considered by many national and international bodies. For example, a UNESCO report has stated that

> countries which are excessively dependent on the import of foreign technologies frequently condemn their national research institutions to play a marginal role in the development process. Furthermore, where the transfer of technology is viewed as an operation affected only by the technical requirements of specific projects, its overall repercussions on the national economy are neither perceived nor evaluated, for example in terms of their social cost. Unfortunately, these repercussions can be much more extensive and durable than the effects initially envisaged for a given development project. Uncoordinated technology-transfer projects can therefore lead to contradictory or even chaotic patterns of scientific and technological development.
>
> A fragmentary approach of this kind has, amongst other things, the drawback of implicitly detracting from efforts to establish an endogenous R&D potential. For if more store is consistently set by imported technologies—as reportedly having more immediate impact on the development process—local research rapidly becomes an abstract exercise, if not a costly option which developing countries can hardly afford (UNESCO 1983).

It maybe recalled that the United States and a few other countries withdrew from UNESCO on the basis of its perceived anti-Western bias, of which the foregoing passage may be an example. Nevertheless, this illustrates a significant and widely held viewpoint, of which the biomedical innovator should at least be aware.

*Cost factors determined by the local administration.* The cost of imported biologicals is inflated in some countries by customs duties that may exceed the initial value of the product. Often, pressure from local industrial and political interests underlie the imposition of such duties. The resulting price differential may favor the purchase of domestic rather than imported products. However, in the case of diagnostic procedures, the cost of reagents represents a relatively small part of overall patient management, and as the consequences of misdiagnosis may be great, other considerations may override cost in the selection of the diagnostic product. The perception of the physician or the laboratory director about the relative quality and reliability of available immunodiagnostic reagents will be an important factor in their decision.

## The Market for Biomedical Innovations in Developing Countries

In the mid-1980s, worldwide expenditure *per capita* for pharmaceutical drugs was US$22. People in developing countries spent an average of $5.4, and in Latin America, $13.8. In the developed world it was $62.1 per capita, and the average U.S. citizen spent $106.3 for pharmaceutical drugs. In 1990 worldwide pharmaceutical sales were estimated at US$130 billion, estimated to grow to US$195 billion by 1995 (Marquez, 1991).

> 77% of the world's population live in developing countries, but only account for 21% of the world's drug consumption. In other words, it is estimated that 1.2 billion people living in developed countries consumed nearly US$95 billion worth of drugs while the remaining 4.1 billion living in developing countries consumed only US$25 billion worth. Of the 5.3 billion people in the world, 1.5 billion have little or no regular access to essential drugs (Anonymous, World Health Statistics Annual 1991).

Vaccines constitute a very small portion of total pharmaceutical sales. An executive of a leading French vaccine manufacturer stated that

> the world's vaccine producers amounted to little more than 1% of the total pharmaceutical industry and the activity was becoming concentrated in an ever-smaller number of firms. At the same time, demand for vaccines had exploded—from around 100 million doses in the early 1980s to 1,200 million doses for 1992. Moreover, within the past few years this tiny industry within an industry was facing a revolution in technology and as a result, in regulatory constraints (Children's Vaccine Initiative 1991b).

Taylor and Laing (1989) estimated world sales of human vaccines at $300 to $400 million, of which (in 1986) $50 to $100 million was spent for the Third World. In 1989 a market research company predicted that the worldwide market for human vaccines would reach $3.7 billion by the year 2000, of which $170 million would be in developing countries (Technology Management Group 1989). It appears, even from these divergent estimates, that the total expenditure for vaccines for the 75% of the world's population in developing countries is less than it is for the 5% of people in the United States. The President's 1992 budget requests for "vaccine-related activities," far more than the cost of the vaccines, was US$495 million, an increase of $35 million (13%) over 1991. The total effective vaccine market in developing countries is, in fact, so small, and risks of all kinds so great, that little commercial effort is devoted to research and development in this field.

> The prospect of recapturing development costs, much less making a profit, is dim indeed if a vaccine offers protection against a disease found almost exclusively in poverty-stricken nations with limited health budgets, staggering national debt, and little hard currency (Freeman and Robbins 1991).

## Defining the Market

An individual with technical knowledge about one aspect of a subject does not necessarily have informed judgment in other areas. It may be natural for a Western academician to presume that because there is a "need," there will be a market, but it is not a simple matter to define the market for a biomedical innovation intended for use in developing countries. As mentioned in chapter 1, benefits may accrue directly to receiving individuals, either those having a particular condition, or those who are to be protected from the condition. In the case of vaccines for the major infectious diseases, young children or others at particular risk, who are actually to receive the inoculation, constitute the epidemiologic target group. In the case of diagnostic reagents, the epidemiologic target may be individual patients, or perhaps persons who are subjects in a large vaccine, drug, or other study.

It is common in developing countries for people to purchase all kinds of drugs over the counter for self-medication, but there is no market for self-medication with vaccines. Demand for vaccines usually occurs only in the face of epidemics, and even then it is for vaccines to be administered by authorities. Individuals may exercise a limited veto by refusing a procedure such as the taking of a blood sample for a diagnostic test or the administration of a vaccine or drug. If enough people were to do so, implementation of the innovation might be prevented in that particular instance. For that reason, public health education or "social marketing" would normally precede or accompany the introduction of an innovation intended for widespread use, particularly where individuals retain some discretion over its application or adoption. Because infants have no opinions about such things, the mother or other caregiver is the client for educational or propaganda messages concerning an immunization program.

*The private sector.* A private physician may prescribe or administer an immunization or pharmaceutical. The product, stocked in his or her medical office, is obtained directly from a pharmacy or wholesaler that orders through conventional business pathways. The producers of a commercially distributed innovation have such routes in mind when thinking of the market for their product.

For diagnostics, the private market in developing countries includes individual physicians who have their own small laboratories, particularly in rural areas. There may be other laboratories in the private sector, either free-standing or attached to proprietary clinics or hospitals. The customer of the laboratory, who makes the decisions about which tests to administer, is the physician rather than the individual patient. But the physician can select only from among the items available. Laboratory administrators, who purchase these items, might import products directly, given sufficient volume. It is more likely that orders will be placed for items already in stock in the country, which will have passed whatever regulatory scrutiny is applied by the government. Depending on distribution arrangements, the product will normally be distributed directly by the manufacturer or by manufacturers' agents or independent distributors who control access by their ordering decisions. "Ultimately, industry executives, not public health officials, make the decisions about which products to develop" (Freeman and Robbins 1991).

*The public sector.* Most countries have (or had) regional or central public health institutes serving as reference laboratories for their respective jurisdictions. Some of these institutes originated in the former colonial days. In francophone countries these were often branches of the Institut Pasteur, and in former British colonies they may have been associated with the colonial or military medical services. Because of the long and intimate relationship between vaccine use and public health, the main laboratory or research institute of the Ministry of Health is often given responsibility for production of certain classes of vaccines and sera. In many cases these have been of inferior quality. An occasional disaster has resulted from use of these products, such as the outbreak of post-vaccinial rabies in Fortaleza, Brazil in 1960 from the use of contaminated locally made vaccine (Pará 1965). The output of these institutes has typically been used within the hospitals, clinics, and health centers of the Ministry of Health.

If new products are intended for use in public health programs and government-operated facilities, decisions about adoption will be made by individuals or committees at the central, regional, or more rarely at the local level. International donor agencies, when involved, may have a say in the decision.

*Who pays for immunization in developing countries?* The market for EPI vaccines has very few paying customers. The actual recipients—the infants and their families—usually do not pay. Wealthier countries may pay from their own resources. In some poorer countries foreign donor agencies, which move vaccines in massive quantities, subsidize most immunization program costs (Table 5–3). Global figures for the Expanded Programme on Immunization have shown that,

TABLE 5–3.   Sources of Financing for National EPI
Programs, Selected Countries, 1979–1987

|  | Distribution of Total Cost (Percent) | |
| --- | --- | --- |
| Country | Ministry of Health | Donor Organizations |
| Routine Strategies | | |
| Turkey | 96.5 | 3.5 |
| Philippines | 84 | 16 |
| Mauritania | 41 | 59 |
| Tanzania | 35 | 65 |
| Burkina Faso | 27 | 73 |
| Campaigns | | |
| Cameroon | 87 | 13 |
| Senegal | 29 | 71 |
| Mauritania | 15 | 85 |
| Mobile Team | | |
| Mauritania | 31 | 69 |

*Source:* Adapted from Brenzel 1990, and the REACH Project.

around 1982, about 79% of the cost for fully immunized children was covered by national sources and 21% from donors, with wide variation among countries (Brenzel 1989). Within the total program (Table 5–4) the cost of the vaccine itself in the countries studied by Brenzel (1990) was borne to a large extent by donor agencies: Cameroon, 38%; Burkina Faso, 59%; Turkey, 60%; Senegal, 99%; Mauritania, 100%.

Informal figures obtained from the EPI in 1992 estimate global donor expenditures on immunization programs at about US$231 million, with multilateral agencies providing 40%, bilateral donors 47%, national UNICEF committees 4%, and other donors 9% (the countries' own contributions are not included). Of the global donor expenditures 26% was for vaccines, 49% for capital investment, 46% for operations and 7% for personnel. The geographic distribution of donor expendi-

TABLE 5–4.  Percentage Distribution of Annual Immunization Costs, by Major Component, Various Countries, Early 1980s

| Country | Cost Component | | | | | |
|---|---|---|---|---|---|---|
| | Salaries[+] | Supervision[&] | Vaccines | Transport | Other | Capital |
| **Ivory Coast** | | | | | | |
| Fixed | 62 | 23 | 10 | 0 | 5 | * |
| Mobile | 24 | 38 | 17 | 8 | 13 | * |
| Kenya | 27 | 43 | 3 | 0 | 11 | 16 |
| Gambia | 38 | 7 | 12 | 15 | 12 | 16 |
| Cameroon | 60 | + | 13 | 8 | 5 | 14 |
| **Ghana** | | | | | | |
| Fixed | 51 | + | 6 | 21 | 7 | 13 |
| Mobile | 51 | + | 4 | 27 | 4 | 14 |
| Indonesia | 25 | 12 | 21 | 14 | 10 | 18 |
| Philippines | 31 | 22 | 12 | 14 | 2 | 19 |
| Thailand | 28 | 20 | 13 | 9 | 4 | 26 |
| **Brazil** | | | | | | |
| Fixed | 68 | 14 | 5 | 1 | 3 | 9 |
| Campaign | 39 | 19 | 4 | 6 | 25[#] | 7 |
| Range | 24–68 | 7–43 | 3–21 | 0–27 | 2–25 | 7–26 |
| Medium | 38/39 | 20 | 10/12 | 8/9 | 5/7 | 14/16 |

*Source:* Adapted from *Social Science Medicine 23*. Creese AL. Cost effectiveness of potential immunization interventions against diarrhoeal disease. Pp. 231–240. Copyright 1986, with kind permission from Pergamon Press Ltd, Headington Hill Hall, Oxford OX3 OBW, UK.

For some countries these data are based on assessments at samples of health units; for others they are disaggregations of national program expenditures.

*included under "Other"

[+] Salaries in most cases refer to payments to immunization personnel. For Cameroon and Ghana salaries include salary component of supervision.

[#] Publicity = 5%

[&] Salary costs of district/regional/national program management, apportioned.

tures (by WHO region) was: Africa 30%, Southeast Asia 29%, Eastern Mediterranean 14%, Americas 13%, Western Pacific 5%, Europe 0%, and global expenditures 19.5%.

The amount of money devoted to global immunization programs appears less formidable when viewed in perspective:

> Vaccines cost money and the people most in need are those least able to pay for them. The cost of routine immunisation in the EPI is about US$10 per head; so the cost of financing the global immunisation programme for one year could be met by cutting the world's defence expenditure by one part in a thousand (Moxon 1990).

*The Vaccine Independence Initiative (VII).* This initiative, supported by WHO and UNICEF, has been strongly endorsed by the Global Advisory Group of the EPI. It

> promotes the sustainability of national immunization programmes by aiding certain countries to ensure the long term supply of WHO approved vaccines through the provision of flexible financing terms, emphasis on strong planning and inter-ministry cooperation and an efficient vaccine procurement system (Expanded Programme on Immunization 1992a).

## Opportunity Risk

There are always at least two choices facing the decision-maker. One is the choice to do nothing. Inaction is not always cheaper, either monetarily or politically. In fact, that may be the most costly alternative.

Where health budgets are level, their allocation is a zero-sum game in which an increase for one program requires a decrease elsewhere. Spending money for health services, particularly for preventive interventions such as immunizations, is commonly justified on the grounds that small current expenditures will avert the need for larger expenditures in the future. Nevertheless, in many developing countries preventive services for poor and rural people often compete unsuccessfully with curative services for the urban elites. Programs considered beneficial, or essential, or even indispensible, may not be offered because it is claimed that there is just no money available for them. Under these circumstances, the introduction of any preventive public health innovation is particularly difficult, even where it can be demonstrated to be cost-beneficial and cost-effective.

## Alternative Risk

This type of risk refers to the existence of entirely different means or procedures that could be substituted to accomplish a similar epidemiologic objective. The type and number of feasible control alternatives varies with the disease and with the local situation. In the days before measles vaccine became available it was usual for health authorities to tack a "Quarantine" sign on the residence of any case, because the only available means of control, and not much control at that, was isolation. Since 1963, measles vaccine has provided the more effective alternative of immunization, both for individual protection and for community control of transmission. Measles vaccine, in a generic sense, has been extremely successful not only because of its efficacy and safety but because there are no alternatives to its use (but see below under "Competitive Risk"). The fact that another viral disease, yellow fever, is transmitted by mosquito vectors expands the range of possible methods for protection and control. In general, the more complex the transmission patterns of the pathogen, the more susceptible it is to attack, and the more levers there are to reduce individual exposure and community distribution of the pathogen. For example, in addition to vaccination, avenues of control might include casefinding and chemotherapy; chemoprophylaxis; control of vectors and reservoir hosts; education; sanitation and water supply; and window screening, bednets, and similar protective measures. Each additional point of vulnerability reduces dependency on any particular strategy (e.g., immunization) and provides opportunities for entirely different kinds of control modalities.

To merit implementation, a proposed technologic innovation must display clear superiority over other options that could achieve the same health goal. The advantages of one procedure over another are generally cast in terms of cost-effectiveness, but may accrue as supplementary benefits, as might happen when a vaccine against one pathogen could protect also against a related organism.

Phase III field trials for safety and efficacy of poliovirus vaccine in 1954 and of measles vaccine a few years later did not require the additional layers of analysis that must now be done before many current biomedical innovations can be considered feasible. Quite apart from the basic biology of the pathogens, the developers of an antiparasite vaccine, for example, have an inherently more arduous task than the developers of those antimeasles and antipolio vaccines. Their product must be not only safe, biologically efficacious, stable, and potent, but the *concept* of immunization must be shown to be superior to other methods that may accomplish similar degrees of personal protection and reduction in transmission.

An extensive review of the literature has shown that vaccine trials typically are conducted with minimal consideration of other existing or potential means of control, in part because few such means exist for most currently vaccinizable diseases. The fact that a candidate vaccine is safe and efficacious is certainly a necessary, but not always sufficient, condition for its adoption in the field. Where alternative means for control are available, the utility of the vaccine must be assessed not only *qua* vaccine, but in comparison with the full range of other control methodologies. The technology assessment is made vastly more complex and costly, but this appears to be unavoidable.

The possibility that an entirely different approach may prove less costly and more effective introduces a high degree of alternative risk: an antiamebic vaccine must be demonstrably better than handwashing or sanitation; an antimalarial or antitrypanosomal vaccine must have clearly specified advantages over vector control, and so forth. In addition, the possibility of harm from an invasive procedure such as immunization is always present, adding to the need for its justification.

In most vectorborne and parasitic diseases, including malaria, a variety of methods to mitigate infection have been known and applied for decades. The classic

15-year study of schistosomiasis control (Jordan 1985) compared three approaches to schistosomiasis control—mollusciciding, water supply and sanitation, and casefinding and chemotherapy—for resident populations on the island of St. Lucia. The possibility of artificial immunization was not even considered.

The definition of target population and epidemiologic rationale is essential for assessment of any innovation. A vaccine intended for community immunization programs must be evaluated alongside these alternative means of control; one intended solely for protection of temporary visitors to an endemic area (tourists, businessmen, military) must demonstrate only safety and short-term efficacy, and need not be compared to alternative control methods.

## Competitive Risk

Market competition will pit the innovation against comparable or similar (or perhaps even identical) products, if any, that could be manufactured or distributed more cost-effectively, be more acceptable to purchasers and users, have better shipping and storage properties, and so forth. The potential appearance at any time of such a product is a continuing element in competitive risk.

An important element in competitive risk is the period of time during which the particular innovation remains competitive. The many companies in the international biomedical marketplace spend millions of dollars annually for research and development of new products, and also adapt published academic research findings. In this technological environment the useful life span of a particular innovation may be only a few years, during which investments in R&D and specialized production requisites must be recovered through economies of scale.

Introduction of an innovation into a new market is always costly to the supplier, including clinical trials, regulatory hurdles, importation or local manufacturing, storage and distribution, a sales team if not already on hand, and so on.

It may be that there has been no previous market for a similar category of product, because (1) such a product did not exist or was not available; or (2) there was no effective demand, because of (a) no locally perceived need for the product, or (b) lack of ability to pay for it.

If an available product already fills an analogous function, then the advantages *to the user* of the innovation will need to exceed a certain threshold in order for its adoption to be considered. Established systems are dislodged with difficulty; novelties are resisted; and organizational inertia and inflexibility do exist, but these may be well grounded. First, civil servants who may be criticized for making an incorrect decision are likely to take no initiatives, and to remain with a safe and familiar, if imperfect, routine. Second, apparent advantages such as time saving or ease of use of a diagnostic procedure, significant in a high-wage Western environment, are of only marginal importance in developing countries where low wages and high staffing levels are the rule. Third, the cost of the procedure must be taken into account, including not only the obvious costs of apparatus and supplies, training of technicians, and so on, but all consequences of the procedure including treatment of those newly found positive by a more sensitive diagnostic test. Fourth, there may be justified anxiety about long-term availability of necessary reagents and supplies. Fifth, there may be well-entrenched elements in favor of maintaining the status quo. For example, relations with existing suppliers may be cordial and long-standing.

*Competitive Risk: The case of measles vaccines.* Measles virus is found throughout the world and represents a major threat to children, particularly where nutrition is poor and other infectious diseases are common. Even in the United States, approximately one case in 2,000 has led to encephalitis, often with permanent brain damage, and about one in 3,000 is fatal. The biology of interactions among the infective agent, host age, maternal antibody status, and vaccines is discussed in Appendix C.

In 1963 both inactivated and live attenuated Edmonston B strain, derived from the original Edmonston-Enders (E-E) vaccines, were licensed in the United States, but the inactivated vaccine was discontinued after a few years. Work on further attenuation of E-E–derived live viruses, primarily in chick embryo fibroblast cultures, was intensive during this period. The resultant vaccines included the Schwartz strain (1965) now used in Europe and many developing countries, and the Moraten strain (1968), long the standard measles vaccine in the United States. The Moraten vaccine does not induce very good immunoconversion in young infants, presumably because of the persistence of maternal measles antibody, and has caused disquiet among public health officials who are always seeking better protection.

Immunization against childhood measles reduced incidence by more than 99% in the United States, so that from 1981 to 1988 only about 3,000 cases per year were reported. The occurrence of several hundred cases of measles in young children in Los Angeles County, California, in 1987 and 1988 was followed by many other outbreaks: the number of cases reported in 1989 was 18,193, and in 1990 the number increased to 27,672 spread among 49 of the 50 states. Of these cases, 22.7% resulted in complications, 21.1% required hospitalization, and 89 children died.

These major measles outbreaks, primarily among unvaccinated preschool-age children, prompted the development of new immunization guidelines. In view of increasing risks of measles infection in the United States, it is important to protect children much earlier than 15 months of age. Therefore the standard recommended age of 15 months for measles immunization in the United States has been reduced to 9 months in officially "high risk" counties, i.e., those reporting at least 5 cases of measles in preschool children during each of the previous 5 years.

The epidemiology of the disease in the U.S. is changing, and the distribution of cases is shifting from older, previously vaccinated, school-age children to younger, unvaccinated children. As more outbreaks occur in younger children, more infants less than 1 year old are exposed. The principal cause of the re-emergence of measles in the U.S. is the failure to vaccinate children at the appropriate age. Although very effective when used properly, the current vaccine has deficiencies as a public health tool. There is a primary failure rate of about five percent, and thus, susceptible individuals accumulate in the population. The failure rate is higher if the current vaccine is given at less than 15 months of age when maternal antibody interferes with vaccine efficacy (NIH, 1992).

More than 10 different derivative live attenuated measles virus vaccines are now used in various countries. The precise basis for the subtle differences among them is difficult to document.

One vaccine strain, the Edmonston-Zagreb (E-Z), was developed in Yugoslavia around 1970 after numerous passages through WI-38 human diploid cells. Comparisons of E-Z with Schwartz-strain vaccines, (e.g., Whittle et al. 1988 in the Gambia; Markowitz et al. 1990 in Mexico; Job et al. 1991 in Haiti) have in general demonstrated that E-Z shows higher immunogenicity (measured as seroconversion and geometric mean titers) when administered in sufficiently high doses containing 10 to 500 times more vaccine particles than standard-potency vaccines. High seroconversion rates were found even in 6- to 9-month-old infants who might still carry some passively transferred maternal antibody. That is precisely the period at which measles is most severe and case-fatality rates are the highest. A preliminary study of vaccine efficacy in Guinea-Bissau (Aaby et al. 1988) suggested that E-Z may be more protective than Schwartz vaccine in an endemic area. Because of these findings, the Global Advisory Group of the Expanded Programme on Immunization recommended in 1989 that E-Z vaccine should be used against measles at 6 months of age. Subsequently, Kiepiela et al. (1991) found that standard-potency E-Z vaccine induced poor serological responses in black children in South Africa, and Garenne et al. (1991) reported from Senegal that children immunized with high-titer E-Z or Schwartz vaccine had a significantly higher mortality rate (from all causes) than those given standard vaccine. Similar reports of higher mortality in E-Z vaccine recipients came also from Guinea-Bissau and Haiti. In response, a WHO Task Force in June, 1992 recommended temporary suspension of the use of E-Z vaccine, and the Global Advisory Board of EPI acted in October, 1992 to confirm the suspension.

The large literature on measles immunization includes numerous recommendations for a variety of age schedules, and points out the complexity of dealing with a relatively straightforward, well-known disease caused by a simple, invariant viral agent.

Despite universal recognition of the seriousness of measles and 25 years of intense experience with well-characterized vaccines, an estimated million children a year still die of measles. The case of smallpox notwithstanding, this reality argues against overestimating the epidemiological impact of a technologic innovation.

*Competitive risk: The case of hepatitis B vaccines.* Hepatitis B is another widespread viral disease. There are an estimated 300,000 annual infections in the United States alone. Infection acquired early in life contributes to the burden of chronic liver disease many decades later. Among the consequences are hepatitis, cirrhosis, and hepatocellular carcinoma. Chronic carriers are found in a proportion estimated at 0.1% to 15% in various populations, totaling hundreds of millions of people around the world. A WHO Technical Advisory Group on Viral Hepatitis (1988) has estimated that more than 40% of persistently infected persons who survive to adulthood will die as a result of the infection.

Hepatitis vaccine is not intended primarily to protect against immediate illness, as with the standard EPI series, but against diseases later in life. It is essentially a vaccine against liver cancer. Infants born to carrier mothers are at high risk of acquiring infection and becoming chronic carriers themselves, but this risk can be reduced by immunization at birth or in early infancy. In Southeast Asia, mothers who carry the hepatitis B virus infect one of every five babies at birth. In many parts of southeast Asia and Africa, hepatitis kills more children than measles (Ryan 1987).

In the case of measles vaccines the various alternative formulations, derived from different attenuated strains of the same virus, are basically similar and share many essential characteristics. For hepatitis B the situation is not so simple. Hepatitis B vaccines in common use are made in two quite different ways. The first is a derivative of pooled human blood plasma (Fagan and Williams 1987). The second is a biotechnologic product synthesized by the yeast *Saccharomyces cerevisae* into which a plasmid has been added that contains the gene for making hepatitis B surface antigen (see Appendix A). Other feasible means of making hepatitis B vaccines include the use of hybrid vaccinia viruses and chemically synthesized peptides (Zuckerman 1985).

By 1989 there were 12 licensed manufacturers of plasma-derived vaccine: five in China, three in Japan, two in Korea, and one each in France and the United States (Maynard et al. 1989), and at least two manufacturers of the recombinant type. Antibody responses and persistence of antibody is similar for both kinds of vaccines. In one small group of people immunized with recombinant vaccine, 32.3% no longer had protective levels of circulating antibody after four years (Jilg et al. 1989). However, residual protection may persist much longer because of an immediate anamnestic response originating from host lymphocytes primed for hepatitis surface antigens.

As discussed in chapter 2, it has long been suggested that hepatitis B vaccine be made a part of the EPI, and the Global Advisory Group has now defined the specific conditions under which this should be done. In some countries, such as Italy, hepatitis B immunization has become mandatory. (DaVilla et al. 1992). The Immunization Practices Advisory Committee of the U.S. Public Health Service and the American Academy of Pediatrics have endorsed a similar policy for the United States (Shapiro and Margolis 1992).

Because of the desperate need for a cheap vaccine, several groups have experimented with low-dose intradermal immunization as a cost-saving strategy. Bryan et al. (1992), reviewing these studies, have concluded that "low-dose intradermal or intramuscular immunization offers protection against hepatitis B at significant savings and may be useful for mass immunization of populations at high risk." Plasma-derived vaccine could be produced for as little as US$0.10 per dose in large quantities (Mahoney 1990). Plasma-derived vaccine has an impressive record of safety and is used routinely in some developing countries, but is no longer produced or used in the United States. Despite assurances of complete protection from possible HIV contamination, it remains unacceptable to many people. Some persons object to accepting any plasma-based vaccines on religious grounds (Milne et al. 1989).

Two yeast-recombinant hepatitis B vaccines made by large manufacturers in the

United States and Europe have become widely adopted and are licensed for use in the United States. The vaccines are effective but remain far too costly for routine application in developing countries.

Hepatitis A virus is spread by fecal contamination of water, milk, or food such as shellfish. Explosive outbreaks sometimes occur in institutions or among military troops. In 1992, an attenuated vaccine against hepatitis A was tested successfully and made available in the United Kingdom, Switzerland, Belgium, Austria, and Ireland (Bancroft 1992; Lancet 1992).

*Competitive risk: The case of poliomyelitis vaccines.* The almost universal use of polio vaccines since the 1960s has essentially eliminated paralytic poliomyelitis caused by wild viruses in the United States, Canada, Australia, Japan, and most of Europe. In the WHO Region of the Americas almost a thousand cases were reported as recently as 1986. The number of cases declined so rapidly that the Pan American Health Organization established the target of elimination of all wild poliovirus by the end of 1990. That was not quite achieved: In 1991 the number of confirmed cases of paralytic polio was less than 10, and of "compatible" cases was a few dozen. As of July, 1993, no case of paralytic poliomyelitis had been reported from the Western Hemisphere in about 21 months.

Although the goal of worldwide eradication of poliomyelitis has been targeted for the year 2000, the Global Advisory Group of EPI has issued some important caveats. First, there is a concern that some countries may postpone polio eradication efforts until the end of the decade. Second, despite the concentrated efforts of groups such as Rotary International, there remain shortages of financial and technical support. The need for vaccine supply and laboratory networks is so acute that "there may be ultimate failure of the poliomyelitis eradication initiative" (Expanded Programme on Immunization 1992a).

The situation with polio vaccines is rather more complicated than with either measles or hepatitis vaccines because of the existence of two quite different vaccines intended for the same purpose: to prevent cases of paralytic polio. These are the injectable original Salk inactivated polio vaccine, or IPV (sometimes called KPV, for killed polio vaccine) and the Sabin live oral vaccine or OPV. The Salk vaccine, introduced in the mid-1950s, was rapidly adopted throughout the developed world, but was replaced in many countries by OPV, which became available in the 1960s.

According to Hovi (1991), IPV has been used exclusively in Iceland, Finland, the Netherlands, Sweden, and some provinces of Canada, and Norway has returned to its use. In France both IPV and OPV are available.

The IPV has been criticized on the basis of cost, lack of mucosal stimulation, and a reported limited duration of immunity. The OPV is faulted because of the need for a cold chain, poor takes in many developing countries, and potential genetic instability leading to reversion to virulence. Almost all recent cases of paralytic polio in the United States have occurred in OPV recipients or their contacts. The advantages and disadvantages of each are shown in more detail in Table 5–5, which is adapted and amplified from Assaad (1979).

There is no shortage of novel and alternative formulations of polio vaccines

TABLE 5–5.  Poliomyelitis Vaccines: Advantages and Disadvantages

*Advantages of Live Poliomyelitis Vaccine*

1. Confers both humoral and intestinal (mucosal) immunity, like natural infection, and does not require continued booster doses.
2. Rapidly infects the alimentary tract, blocking spread of wild virus.
3. Antibody is induced very quickly in a large proportion of vaccinees.
4. Oral administration is more acceptable to vaccinees than injection, easier to accomplish, and negates the risk of AIDS transmission from reused needles and syringes.
5. Administration does not require the use of highly trained personnel.
6. When stabilized with magnesium chloride, OPV can retain potency under field conditions with little refrigeration.
7. Immunity induced may be lifelong.
8. Relatively inexpensive to produce and to administer.

*Disadvantages of Live Poliomyelitis Vaccine*

1. The vaccine viruses can mutate, revert to neurovirulence, and (rarely) cause paralytic poliomyelitis in recipients or their contacts.
2. Vaccine progeny virus spreads to household contacts. [Note: Some people consider this spread to be an advantage, but the progeny virus spread by vaccinees may have mutated, with unknown characteristics.]
3. Vaccine progeny virus may spread to persons in the community who have not agreed to be vaccinated.
4. The vaccine has been very successful in the industrialized countries but induction of antibodies has been relatively difficult to accomplish in the tropics unless repeated doses are administered. In some areas, even repeated administration has not been effective.
5. Contraindicated in those with immunodeficiency diseases and their household associates, as well as in persons undergoing immunosuppressive therapy.
6. The virus is relatively stable at refrigerator temperatures, but loses its efficacy if exposed to heat. Can be stabilized to some extent with magnesium chloride, but a continuous cold chain is still needed from manufacture through use, to prevent inactivation of the viruses. This remains difficult to accomplish in many areas.
7. The three serotypes interfere with each other and all may not "take" after the first dose. The vaccine may also interfere with other live viral vaccines.
8. Type 2 vaccine virus may interfere with responses to types 1 and 3, and the absolute and relative dosages of the three components may need to be adjusted locally from time to time.

*Advantages of Inactivated Poliomyelitis Vaccine*

1. Confers humoral immunity in a satisfactory proportion of vaccinees if a sufficient number of doses is given.
2. Can be incorporated into regular pediatric immunization with other vaccines (DPT).
3. Absence of living virus precludes potential mutation and reversion to virulence. Therefore this vaccine may be used in the final stages of an eradication campaign in which a small number of cases might be caused by reversion of the attenuated viruses in OPV.
4. Absence of living virus permits its use in immunodeficient or immunosuppressed individuals and households.
5. Appears to have greatly reduced the spread of polioviruses in small countries where it has been properly used (wide and frequent coverage).
6. May prove useful in certain tropical areas where live vaccine has failed to take in young infants.

continued

151

TABLE 5–5.   Poliomyelitis Vaccines: Advantages and Disadvantages *(continued)*

*Disadvantages of Inactivated Poliomyelitis Vaccine*

1. Does not induce local (intestinal or mucosal) immunity in the vaccinee.
2. More costly than live vaccine.
3. Use of virulent polioviruses as vaccine seed could lead to a tragedy if a failure in virus inactivation were to occur.

*Source:* Adapted from Assaad 1979.

(Patriarca et al. 1988, McBean et al. 1988), including an inactivated formulation of the trivalent live attenuated vaccine. An enhanced potency IPV, often called E-IPV, was introduced in the mid-1980s, and genetically altered live poliovirus recombinants have been proposed by a number of authors.

An attractive program combining the good features of both types of vaccine was described by Hovi (1991):

> To solve the problem of the vaccine-associated paralytic disease due to OPV on the one hand and the poor mucosal immunity obtained with IPV on the other, it may not be necessary to develop new recombinant viruses but rather to use existing vaccines in an optimized way. . . . The frequency of VAP [vaccine associated paralysis] recorded after giving OPV to persons with a previous history of antipolio immunization is vastly lower than that after the first dose of OPV. Thus, the safest way to obtain the good mucosal immunity associated with live virus vaccination would be the one adopted in Denmark, i.e., administer 2 or 3 injections of IPV to infants, followed by boosters of OPV. This mode of anti-polio immunization has been followed in Denmark since 1967 with very good results.

Beale (1990) has reviewed the relevant data, pro and con, and suggested three policy options:

1. Persist with OPV for infant immunization and as the basis of a global eradication program. This option would accept the small number of vaccine-associated cases and lowered "take rates" in developing countries, and would require additional research on improved attenuation and vaccine delivery systems.
2. Adopt IPV in place of OPV. According to Beale (1990) "the case for a change to IPV seems irresistible, but it clearly lacks general support." Problems include (a) the fact that success depends on high immunization rates, which could be achieved through combinations such as DPTP; (b) the fear, perhaps unfounded, that herd immunity would suffer; and (c) availability and cost of the vaccine.
3. Use IPV and OPV sequentially. This is the policy in Denmark. Each country or jurisdiction must decide which plan to use on the basis of its own interests, including cost.

The Global Advisory and Technical Consultative Groups of EPI have considered all the data and concluded that OPV remains the vaccine of choice. The switch to IPV as an end-game strategy is recognized but has not been made official policy.

## Production

As the product moves along the research and development steps, increasing resources, both in terms of manpower and investment, will be required. Based on our own experience with a recombinant vaccine, the research phase can represent as little as 10% of resources needed to develop a project (DeWilde 1987).

### Scale of Production

Biologicals are among the most difficult items for most pharmaceutical companies to produce reliably, uniformly, and profitably. In comparison with the chemical synthesis and formulation of conventional tablets, capsules, and syrups, the manufacture of biologicals such as immunodiagnostic reagents and vaccines presents complex and sometimes daunting technical obstacles. It is one thing to coat a few dozen microtiter plates or make a few milliliters of vaccine for a research project, and quite another to establish reliable procedures for manufacturing ten thousand identically coated plates of hundreds of liters of highly standardized vaccine every week. Scaling up to pilot plant quantities, and then to production lots, are problems for which the typical biomedical research scientist is untrained and unprepared. As additional genetically engineered products appear on the market, experience is being gained and production processes are becoming more rigidly controlled.

### Stages of Quality Control in Pharmaceuticals

The oldest pharmacopoeias tried to assure quality by defining the raw materials and the manufacturing process as precisely as possible. Then, when alternative production methods were developed, emphasis shifted to normative specifications and the development of sensitive methods to detect impurities in the final product: "anything that need not be there should not be there" (van Noordwijk 1988). Now it is realized that quality must be built into the entire process of manufacture through stringent adherence to Good Manufacturing Practices (GMP).

Relying on experience gained from industrial-scale production of recombinant hepatitis-B vaccine, DeWilde (1987) reemphasized the need for consistency of the process and product. Besides the usual considerations of purity and efficacy, the production of antigens by genetic engineering methods generates special concerns, including (1) the genetic stability of the host and vector; (2) fidelity of expression, and avoidance of translational errors or mutations in the high-level expression and large-scale culture systems; and (3) verification of identity with the natural product. As generic examples of quality-control standards, Table 5–6 provides guidelines for production of biologicals.

### The Meaning of Quality

The question that tormented Robert Pirsig (1974)—*What is quality?*—becomes foremost when the time arrives for pilot production, clinical trials, and full-scale production. Assurance of the quality of biologics depends on two elements: (1)

TABLE 5–6.    Favorable Conditions for Biologicals

---

• Lyophilization, with optimal residual moisture and with a matrix structure provided by excipients such as gelatin
• Freezing, but with maintenance of constant temperature to prevent disruption by shifts in ice crystals
• Purity, especially with freedom from destructive enzymes (proteinase, lipase, amylase)
• Freedom from natural or added surfactants
• Freedom from atmospheric oxygen and other oxidants
• Addition of "protective" substances such as proteins, polypeptides, amino acids, sugars, glycerides, etc.
• Optimal pH
• Optimal cationic environment, especially in relation to enzymic activities (chelators or added salts)
• Cross-linking of amino groups by formaldehyde to stabilize protein configurations
• Darkness and freedom from photoactive substances

---

Source: Hilleman 1989.

definition and specification of the relevant criteria, and (2) quality control of production and utilization.

> Quality control [of drugs] exists in most countries. More than 70% of the 104 countries reviewed have some mechanism for assessing the quality of products or have a quality-control laboratory. However, in many of them, this control does not function effectively (Anonymous, World Health Statistics Annual 1991).

In the United States, standards for biological products including commercially distributed immunology and microbiology devices and reagents are governed under the Code of Federal Regulations, Title 21, chapter 1, subchapter F, that covers the Food and Drug Administration. Section 610 describes general requirements of safety, potency, sterility, purity, and so forth. More specific criteria are defined in sections 660 and 866. Dozens of serological reagents for viral, bacterial, and parasitic pathogens are specified. Although written for the United States, these standards have become internationally recognized as the basis for regulation and as a focus for technical "harmonization activities" by multinational agencies such as the European Economic Community and the World Health Organization.

In the United States, government regulations for biologics have been extended to monitor the safety, purity, and potency of new drugs and biologicals produced by recombinant DNA technology. These regulations require collection and maintenance of data on the consistency of yield of the product from full-scale culture, with criteria established a priori for the rejection of culture lots. Biologics made by recombinant DNA technology must not contain detectable viruses, nucleic acids, or antigenic materials.

Quality control is expensive. It accounts for 90% of the production cost of Sabin attenuated polio virus vaccine grown on Vero cells (Dupuy and Freidel 1990).

*Concerns about vaccine quality.* Vaccine supply is a priority area for the decade of the 1990s. As demand increases and additional vaccines such as hepatitis

B are incorporated into immunization programs, costs increase and production capacity is being strained.

> There is also concern about the quality of vaccines. An increasing number of incidents are being reported where vaccines have been found to be of poor quality or where infrastructure at the country level is insufficient to assure the maintenance of the relevant WHO standards of quality. Production in developing countries must be based on the proper transfer of technology and the strengthening of national control authorities to monitor quality and good manufacturing practice (Expanded Programme on Immunization 1992b).

A specific example of quality-control problems is in the use of a group B meningococcal outer membrane protein vaccine produced by the Ministry of Health in Cuba. This vaccine has reported efficacy of 83% in Cuba, but when used in children in São Paulo, Brazil, its efficacy was estimated at only 40% to 50%. The Steering Committee for Encapsulated Bacterial Vaccines of the Programme for Vaccine Development (see chapter 6) concluded that "the composition of the Cuban vaccine may be difficult to characterize and lot-to-lot variation could explain the discrepancy [in] results between Cuba and São Paulo" (Programme for Vaccine Development 1991).

Differences among producers can account for variations within different vaccines distributed under the same name. For example, Galazka (1989) showed that after 28 days at 37°C the viability of batches of BCG ranged from 66% to 3% depending on the manufacturer and substrain used. Similar differences may occur among batches of yellow fever vaccines.

## Quality Control and Technical Support

Biologicals and other biomedical innovations must be produced in accordance with the principles of Good Manufacturing Practices (GMP) as defined in published industry and governmental standards. Adherence to these standards requires a large investment in equipment, supplies, personnel, and time. Persons experienced in industrial production of biologics recommend that for each worker engaged in manufacturing there be one person working exclusively on quality control.

The characteristics of biologicals are affected by many conditions of production. The quality and uniformity of raw materials, including amino acids, peptides, hormones, salts, vitamins, water, serum, and other components of culture media, and even of the plastic and glassware in which they are stored and used, must be absolutely consistent. The same is true of temperature, gas phase, and other characteristics of the culture or fermentation environment. Even so, hybridomas may suddenly stop producing antibodies, or produce at lower yields, or change other characteristics in response to subtle or uncontrollable variations in conditions.

*Biohazards.* Research personnel, production workers, and local inhabitants must be protected from potentially damaging organisms and procedures. The same is true of the surrounding environment and biota. Therefore adequate containment facilities must be constructed to prevent the release of infectious materials in wastes or through accidents.

*Shelf life.* The scale of production is linked closely to the supply of raw materials and demand for the product. Timely production is particularly important in view of the relatively limited shelf life for many biotechnologic and innovative products such as diagnostic reagents and vaccines. Attenuated or recombinant living vaccines may have an absolute requirement for continuous refrigeration (the "cold chain") within closely specified limits. Enzymes and other products may need to remain frozen at suitable temperatures. In situations where power supplies are unreliable, adequate storage may be difficult, necessitating frequent small shipments to the point of use. Maintenance of a suitable temperature during distribution to remote localities may present a logistic nightmare, particularly during periods of large-scale mass vaccination campaigns where stocks must be sent simultaneously to many sites. The stability of EPI vaccines at various temperatures is shown in Table 5–7.

## Quality Assurance in Utilization

However well a product is manufactured, shipped and stored, its ultimate value depends on the way it is used. Poor technique and inadequate management will cause even the best products to perform below their design potential. For diagnostics, quality assurance is a way of life in well-run clinical laboratories.

The withdrawal of samples of vaccine from the field for potency testing is often advocated, but this is not so simple a matter. The vaccine may need to travel great distances and for long periods of time to a laboratory where such work can be done, and if found deficient it may still not be certain whether the vaccine had been potent at the point of use. Moreover, potency testing is a complicated and costly procedure and results may not be known for months after the sample is taken, making the whole procedure less valuable. Galazka (1989) has pointed out that only a large number of doses (from 2,000 for measles to 200,000 for DPT) justifies sending a routine vaccine for retesting. (Such figures do not refer to experimental uses such as clinical trials.) Proficiency of laboratory workers must be evaluated from time to time by use of coded specimens of known characteristics. Requisites for quality assurance in utilization are listed in Table 5–8.

Greenwood and Whittle (1981d) specified a number of common reasons for failure of immunization schemes in the tropics. They lament that

> over-heating may occur at the airport, where vaccines may be stored without refrigeration for long periods, whilst complex negotiations with customs authorities take place. On one occasion we found a large consignment of rabies vaccine in an airport refrigerator which was not working; the vaccine had been there for several months as the recipient had not been informed of its impending arrival. . . . Power-failures are a feature of every-day life in many parts of the tropics and vaccines

TABLE 5–7. Stability of EPI Vaccines at Various Temperatures

| Vaccine | Storage Temperature in °C | | | |
| --- | --- | --- | --- | --- |
| | 0–8 | 22–25 | 35–37 | over 37 |
| Tetanus and diphtheria toxoids as monovalent vaccines or components of combined vaccines | Stable for 3–7 years | Stable for months | Stable for at least 6 weeks | Stable for 2 weeks at 45°C, loss of potency after few days at 53°C and after few hours at 60 to 65°C |
| Pertussis vaccine | Safe storage for 18–24 months although with continuous slow decrease of potency | Stability varies. Some vaccines stable for two weeks | Stability varies. Some vaccines lose 50% of potency after one week storage | At 45°C about 10% lose of potency per day, rapid loss in potency at 50°C |
| Freeze-dried BCG vaccine | Stable for one year | Stability varies. 20–30% loss of viability after 3-month storage | Stability varies. 20% loss of viability after 3–14 day storage | Unstable. 50% loss after 30 minute exposure to 70°C |
| Reconstituted BCG vaccine | Reconstituted BCG vaccine should not be used during more than one working session (5–6 hours). This recommendation has two bases: (1) concern over the risk of contamination, as BCG contains no bacteriostatic agents, and (2) concern over the loss of potency. | | | |
| Freeze-dried measles vaccine | Stable for two years | Retains satisfactory potency (half life) for one month | Retains satisfactory potency for at least one week | 50% loss in potency after 2–3 day exposure to 41°C. At 54°C 80% loss in potency after one day exposure |
| Reconstituted measles vaccine | Unstable. Should be used in one working session | Unstable. 50% loss after one hour. 70% loss after 3 hours | Very unstable. Titre may be below acceptable level after 2–7 hours | Inactivation within one hour |
| Oral vaccine poliomyelitis | Stable for 6–12 months | Unstable. 50% loss after 20 days. Some vaccines may retain satisfactory titers for 1–2 weeks | Very unstable. Loss of satisfactory titer after 1–3 days | Very unstable. At 41°C 50% loss after one day. At 50°C loss of satisfactory titer after 1–3 hours |
| Inactivated poliomyelitis vaccine | Stable for 1–4 years | Decline of D-antigen content for type I after 20 days | Loss of D-antigen for type I after 20 days in some vaccines but no significant loss after 4 weeks in others | Precise data lacking |

Source: Galazka 1989.

157

TABLE 5–8.   Requisites for Quality Assurance in Utilization
of Immunodiagnostic Procedures

- Adequate preliminary research and development of the product
- Adherence to Good Manufacturing Practices
- Proper packaging, shipping, storage, and use before the expiration date
- Suitably trained and motivated personnel
- Supportive administrative and institutional framework
- Refresher and in-service courses to maintain and update skills
- Proper collection, preservation, shipping, and handling of specimens
- Use of appropriate methods
- Suitable design, construction, and maintenance of laboratory space
- Testing of components such as media, stains, and reagents
- Maintaining equipment in good repair and proper calibration
- Using appropriate controls and standards
- Maintaining proper records
- Proficiency testing of personnel by coded control specimens
- Use of split or duplicate samples for consistency testing
- Comparing results with other laboratories
- Communication with the persons who sent the specimens and with those who will act on the diagnosis
- Acting to correct lapses and departures from standards of quality

TABLE 5–9.   Percentage of Measles Seroconversion at
Vaccination Centers, by Time of Vaccine Use after
Dilution, Republic of Cameroon, 1970–1971

| Duration of Vaccine Utilization | | Percent Seroconversion |
|---|---|---|
| Span | Mean | |
| within 2 hours | 1 hr 13 min | 44.0 |
| within 3 hours | 1 hr 58 min | 38.3 |
| more than 3 hr | 3 hr 10 min | 28.6 |

Source: Adapted from McBean et al. 1976.

may be damaged whilst being held by central or local health authorities. . . . It is surprising that few health authorities check the potency of their imported vaccines before starting on a major immunization campaign. . . . A spot survey carried out during a measles campaign in Nigeria showed that 19 of 20 reconstituted samples of measles vaccine were inactive.

An example of the decline in potency of diluted vaccine is given in Table 5–9 from the classic study of McBean et al. (1976) in Cameroon. Onoja et al. (1992) found that measles seroconversion rates in two health centers in Ibadan, Nigeria, were 64% and 26%, associated with differences in vaccine handling between the two sites, showing that certain problems have persisted over the years since McBean's study.

## Monitoring the Consequences of the Innovation

### *Adverse Effects of Vaccines (Injury Risk)*

The usually minor adverse effects of vaccines are accepted by most people as being preferable to the disease itself. The margin between unwelcome but tolerable side effects and actual physical harm is difficult to define. Bellanti et al. (1987) and Wassilak and Sokhey (1991) have reviewed adverse reactions to vaccines in some detail. A monitoring system for adverse events following immunization ("postmarketing surveillance") has been described by Stetler et al. (1987).

Vaccines can cause injuries in several ways. They may be defective through contamination, adulteration, or other errors in manufacturing or labeling. Even with good, potent vaccine, injury can arise from improper administration or dosage, or when the immunization was contraindicated because of age, allergy, illness, or immunosuppression. In addition, nondefective vaccines properly administered can "produce unforeseeable adverse reactions in individual cases. Although such reactions are not necessarily anyone's 'fault,' any resulting injury is obviously a serious matter to the injured person" (Mariner 1987).

Some make a distinction between harm caused by a vaccine undergoing a field trial and harm caused by a licensed vaccine used for routine immunizations. In the former case, the process of informed consent requires ideally that each vaccine recipient (or responsible guardian) (1) understand the experimental nature and risks of the procedure; (2) can, without prejudice, decline to participate; and (3) agree voluntarily to become a subject in the trial. Where vaccines are in routine use, children may be required to submit to a series of immunizations in order to receive certain benefits from the state, such as enrollment in school. In the latter case persons are under legal compulsion, and may avoid immunization only through considerable effort such as documented religious objection. A person who agrees to become an experimental subject, or to accept a routine immunization, does not thereby waive his or her rights to compensation for any ensuing injury.

*Public perception of immunizations.* The impression among the public that a vaccine is harmful or useless can arise at any time, may lead to community resistance, and can hamper or cripple an immunization program. Often such an opinion is the result of panic or rumor, but sometimes such a perception is understandable. For example, naturally acquired illnesses that should have been prevented can follow the use of impotent vaccines arising from improper storage (including a break in the cold chain for live virus vaccines) or maintenance beyond the expiration date. Occasionally, as described below, outbreaks of disease can occur in populations with high rates of adequate immunization. It is possible that some individuals with preclinical or inapparent infections at the time of immunization may later develop clinical disease, resulting in public outcry. Such a situation was documented for a leprosy vaccine, in which persons harboring the leprosy bacillus near their nerve endings may develop neurological damage because cell-mediated immune response is accelerated (Bloom and Mehra 1984). The provocation of paralytic poliomyelitis by injections is also known (see later).

Confusion may arise in the case of clinical syndromes having varied etiologies. For example, otitis media in children can result from infection with *Haemophilus influenzae, Streptococcus pneumoniae,* or several other agents. A vaccine might give excellent protection against one of these, but children might still come down with otitis media caused by a different pathogen. The distinction is perfectly understandable in scientific terms, but may not be so simple to explain to a mother whose child was immunized and then got an ear infection anyway.

The withdrawal of public cooperation augurs the premature termination of a vaccine trial. The best preventative for such an event appears to be scrupulous quality control of the vaccine itself and of its administration, and a careful program of public education, information, and continuous communication.

## Vaccine Failure (Epidemiologic Risk)

Many instances have been reported of outbreaks of certain diseases in individuals and populations thought to be adequately protected against them. Incidents of vaccine failure are customarily classified as *primary,* in which the vaccine does not induce sufficient protective immunity, and *secondary,* in which immunity is of shorter duration than expected and the immunized person is prematurely exposed to risk of the disease. When vaccine failure occurs, "the community is the first to suffer and with this the credibility of the health services" (DeQuadros et al. 1992).

Part of the pathway leading to primary vaccine failure may be related to varying conditions in different populations and areas of the world. The example of varying responses to *Haemophilus influenzae* type b polysaccharide vaccine among different ethnic groups has already been discussed. In the case of oral poliovirus types 1 and 3 vaccine, seroconversion rates approach 100% in industrialized countries but average only 73% (range, 36% to 99%) in developing countries (Patriarca et al. 1991). Not enough is known about the biological responses to vaccines among different kinds of children.

Measles outbreaks have occurred frequently in presumably well-immunized populations. An example of primary vaccine failure on a grand scale occurred in Hungary during 1988 and 1989. Almost 18,000 cases of measles were reported, mostly in vaccinated people. It was found that those individual vaccinated during the April and September 1973 campaigns were at highest risk of measles. Data suggested that the L-16 vaccine used in 1973 had a low efficacy, was more thermolabile, and may have been compromised during transportation or by use of multi-dose ampoules (Agócs et al. 1992). In another example, measles vaccine produced in Mexico was found to have low efficacy due to rapid loss of potency. When an inadequate stabilizer was found responsible, production of the vaccine was halted. The formulation of vaccine stabilizers was a commercial trade secret, available only at unaffordable cost, whereupon Mexican researchers developed a new and effective stabilizer (Martínez-Palomo et al. 1992). In Harare, Zimbabwe, 4357 cases of measles were seen at primary health care centers in a 1988 outbreak, despite reported vaccine coverage of 83%. Of all cases aged 9 to 59 months, 59% had documented evidence of immunization. About one third of the cases required hospitalization. Kambarami et al. (1991) suggested that under conditions prevailing in Harare, "the failure rate for the standard Schwartz measles vaccine also appears to be high."

Outbreaks of paralytic poliomyelitis have been reported in populations well immunized with trivalent oral polio vaccine (Patriarca et al. 1991). In Oman, a recent outbreak of poliomyelitis had one of the highest attack rates of paralytic disease that has been reported during the vaccine era, with transmission lasting for more than 12 months. Among the most disturbing features of this outbreak was that it occurred in the face of a model immunization programme and that widespread transmission had occurred in a sparsely populated, predominantly rural setting (Sutter et al. 1991).

Epidemiological investigation will distinguish instances of vaccine failure from cases that occur in the small proportion of unimmunized people within a highly protected population (indicating a lack of herd immunity) as happened in Taiwan (Kim-Farley et al. 1984). Patriarca et al. (1988) suggested that vaccine formulation, i.e., the relative ratio of the three poliovirus types, may have played a role in an outbreak in Brazil.

Vaccine failure is difficult to discern in noncommunicable diseases such as tetanus that occur sporadically, rather than in outbreaks. Homma and Knous (1992) have reported that tetanus toxoid made in developing countries (which accounts for 60% of world requirements) does not consistently meet acceptable standards for efficacy.

## Failure Owing to Improper Manufacture of Vaccines

Adverse reactions may result from contaminants in the manufacturing process. The 17D vaccine for yellow fever is grown in chick embryos, and avian leucosis viruses are often present in eggs from nonmonitored flocks. Although there is no epidemiological evidence for its occurrence, there is at least the theoretical possibility of cancer in recipients of such vaccines. Mention has been made of the calamity in Brazil, when 18 fatal cases of postvaccinial rabies were induced by improperly manufactured vaccine containing virulent virus (Pará 1965). Fortunately such occurrences are extremely rare.

*The Lübeck disaster.* The live bacterial antituberculosis vaccine BCG was derived from a virulent strain of *Mycobacterium bovis* isolated in 1904 from a cow with tuberculous mastitis. Calmette and Guérin worked tirelessly with this strain and after more than 230 laboratory transfers, an avirulent, attenuated strain emerged. In 1921 a newborn child was experimentally inoculated with this strain, without adverse effects, and by 1924 BCG (named after the two French experimenters) was used in Europe, Japan, and Brazil (Crispen 1989). In France alone, 242,500 children were immunized safely with BCG between 1924 and May of 1930 (Anonymous 1930). In the intervening decades BCG has been administered to thousands of millions of children.

In 1930 the further use of BCG was jeopardized because of an incident that has come to be known as the "Lübeck Disaster." Although accounts vary slightly, it is clear that around the end of February of that year approximately 251 newborns in the German city of Lübeck were vaccinated orally with a bacterial culture considered to be BCG. Beginning in late April more than 200 of these children showed symptoms of tuberculosis and 75 died. The same BCG culture, which had been originally provided by the Institut Pasteur in Paris in 1929, had previously been used to immunize more than 3,000 children without incident (Anonymous 1930), and was used safely in Riga, Latvia (Kirchenstein, 1930). On investigation it was evident that a laboratory strain of virulent *Mycobacterium tuberculosis* had been either substituted for, or accidentally mixed with the BCG culture by the workers in Lübeck (Youmans 1979, Crispen 1989). Such laboratory errors are far less likely to occur today but the Lübeck disaster of more than 60 years ago points out the need for constant vigilance.

*Polio vaccine and the putative origin of AIDS.* In the 1950s poliomyelitis virus for both live and inactivated vaccines was grown on monkey kidney cells:

> It was found after some years that the kidneys were naturally infected with a virus from the kidney cells which came to be known as SV or simian virus 40. Many early batches of polio vaccine were contaminated with this now-classic oncogenic virus. Tens of millions of people were immunized with both polio and SV40 viruses. Fortunately, no harmful effect of the SV40 in humans has been detected in a number of follow-up surveys (Shah and Nathanson 1976; Mortimer et al. 1981). But we know now that rhesus monkeys, at least in some U.S. colonies, are infected with simian immunodeficiency virus or SIV, which produces effects in these monkeys similar to those produced in humans by HIV. What if the kidney cell cultures had been contaminated with SIV, then unknown, and what if it had been infective to man, and took 5–10 years to show an effect? (Basch 1990).

Several other accounts suggested possible links among monkey kidney cells, polio vaccines, and the origin of AIDS. Kyle (1992) proposed that in the mid-1970s homosexual men took live poliovaccine as a treatment for herpesvirus infections. An article in Rolling Stone magazine (Curtis 1992) generated enough interest to stimulate the creation of a scientific panel to evaluate this theory (Touchette 1992a). In October, 1992 the panel, convened by the Wistar Institute of Philadelphia, concluded that there was no association between polio vaccine and the origin of AIDS. Nevertheless, the panel warned against continuing use of monkey tissue in the preparation of vaccines for human use (Touchette 1992b). The inference of a connection between monkey tissue, polio vaccine, and AIDS, although very likely incorrect, is a useful reminder of the potential for unforeseen complications in any invasive procedure using biological materials.

## Failure Owing to Improper Use of Vaccines

*Potential contamination of vaccines during usage.* Insofar as possible, vaccines should be packaged as individual, nonreusable doses, or protocols must be established to prevent reentry into stock bottles of used syringes or needles. It has been shown in Canada (Langille et al. 1988) that even under presumed high standards of sanitation, faulty practices permitted approximately one-third of previously entered vials of xylocaine to contain debris, including red blood cells and epithelial cells; and almost one-fifth to contain viable microorganisms. Considering the possibility of transmission of HBV and HIV, a similar situation cannot be permitted to develop with respect to vaccines.

*Improper dosage.* In the case of yellow fever vaccine, Brés (1979) reported that improper dosage implies risk. Data show that the 17D live attenuated virus multiplies in the individual within three to seven days, with neutralizing antibodies appearing between 7 and 10 days. When the vaccine dose is too low, viremia is retarded and the delay may be linked to encephalitic conditions. Certain strains of the virus are more neurotropic than others, leading to the WHO requirement that all seed lots be tested for neurotropism in monkeys.

Although current evidence suggests that tetanus toxoid can be administered to pregnant women at any stage of pregnancy without risk of congenital defects or abortions, the rate of local reactions (erythema, pain, swelling) increases with the number of doses received.

*Difficulties encountered in assessment.* Short-range, relatively common complications of vaccines are often encountered in clinical trials, but longer term or more infrequent adverse effects of vaccines may not appear until later, when the vaccine has already been used extensively. Systems of vaccine evaluation must be developed to follow these adverse reactions that are less frequent and less clearly connected to the time of administration of the vaccine. As more powerful immunizing agents are incorporated into vaccines and particularly where transformed cells are used as vaccine substrates, researchers will need to follow carefully the incidence of long-term adverse effects.

*Provocation of paralytic polio by injections.* The likelihood that paralytic poliomyelitis can be provoked by injections has been considered many times since first suggested in 1909 (history in Sutter et al. 1992). The validity of this theory was demonstrated during an investigation into the 1988–1989 outbreak in the Sultanate of Oman, when 118 cases of paralytic polio were reported (Sutter et al. 1991). A subsequent study showed that about 43% of cases had received DTP injections within 30 days before onset of paralysis, while only 28% of age- and residence-matched control children had received DTP in the same interval. Moreover, the anatomic site of the injection corresponded with the anatomic site of paralysis in all 19 children for whom such information was available. Sutter et al. (1992) concluded that the proportion of paralytic cases attributable to DTP injection was 25% for children under two and 35% for children 5 to 11 months of age.

Although there was no evidence that the live polio vaccine virus itself increased the risk of paralysis in this group of children, it is possible that oral polio vaccine given simultaneously with DTP may play a subtle role in the process. Therefore the most prudent course for public health officials is to avoid giving injectable vaccines in the course of oral polio vaccination campaigns. However, it may be difficult to resist the temptation to do, as access to a large assemblage of young children of immunizable age does not arise very often. The potential for provoked cases of polio must be balanced against the expected number of illnesses and deaths that may occur if other vaccines are not given. The best strategy may lie in close adherence to EPI recommendations for vaccine administration at specified ages, and greater efforts at eradication of wild polio virus (Sutter et al. 1992).

## Adverse Reactions: the Case of Pertussis Vaccine

Many authors have commented on adverse reactions to the classical whole-cell *Bordetella pertussis* vaccine, either alone or as a component of DPT. The incidence of certain adverse reactions following immunization against pertussis, and actual infection with *B. pertussis* were compared by Galazka et al. (1984) as shown in Table 5–10. This table indicates that the risk of the various conditions is far higher following natural infection. However, infection is a rare event in a population, and immunization is very common. Therefore the number of actual cases of adverse reactions after infection and after immunization may be much more similar than Table 5–10 may suggest.

After almost 30 years of pertussis immunization in Sweden, and despite high vaccine coverage, a major outbreak with about 800 cases a month occurred between September, 1977 and the end of 1978. Pertussis vaccine was then withdrawn from use in 1979. Between 1980 and 1985 there were two large outbreaks with almost 37,000 confirmed cases, with a cumulative incidence rate estimated at 16% of the unimmunized cohort born in 1980 compared with 5% of the immunized cohort born in 1978.

## Liability for Injury Caused by Vaccine

Kitch (1986) pointed out that resolving the conflict between the society's health and the individual's risk of injury involves both science and social policy, and

TABLE 5–10.    Extimated Rates of Adverse Reactions following DPT Immunization Compared to Complications of Natural Whooping Cough

| *Adverse Reaction* | *Whooping Cough Complication Rate/100,000 Cases* | *DPT Vaccine Adverse Reaction Rate/100,000 Vaccinees* |
|---|---|---|
| Permanent brain damage | 600–2000 | 0.2–0.6 |
| Death | 100–4000 | 0.2 |
| Encephalopathy/encephalitis | 90–4000 | 0.1–3.0 |
| Convulsions | 600–8000 | 0.3–90.0 |
| Shock | 600–8000 | 0.5–30.0 |

*Source:* Adapted from Galazka et al. 1984.

that an attempt to require science to provide a perfectly safe vaccine is unrealistic. In *Reyes vs. Wyeth Laboratories* (1974), the Fifth Circuit Court of Appeals found Wyeth laboratories liable for the injury of a child who had received OPV, even though the vaccine was properly produced and administered, no causal relationship was established, and evidence showed that the infection resulted not from the immunization but from a wild virus. The judge presiding over the case declared that such compensatory measures should be borne by the manufacturer as a "foreseeable" cost of manufacturing vaccines. Such decisions, says Kitch, have created an unstable environment, which has encouraged more claims and has forced manufacturers to pour profits into insuring and defending their products, with a resulting rise in the cost of vaccines.

More importantly, such a climate has discouraged innovation and a renewal of efforts in research and development of new and better vaccines. Because it is not possible to predict exactly how a vaccine will behave in every individual, manufacturers are forced to incur all the costs of research, development, production, and litigation before mass administration, and are liable even for the sort of defects that may appear only once in every three or four million persons immunized. Ironically, those who might request compensation for injury are those most likely to be protected by the vaccine which injured them: "If persons who contract polio from the vaccine are especially susceptible to polio virus, then they are greatly helped, not hurt, by the pervasive use of Sabin polio vaccine" (Kitch 1986).

As the issue of compensation for presumed injury from immunizations became acute in the United States, many manufacturers abandoned vaccine production and distribution rather than be exposed to the risk of potentially devastating litigation. In the mid-1970s epidemiologists at the CDC predicted a major epidemic of influenza and recommended universal immunization in the United States.

> Insurers, wary of legal exposure, began withdrawing coverage from manufacturers, causing them to reconsider continued production. The federal government responded by passing the Swine Flu Immunization Act, under which the government for the first time accepted liability for injuries arising from an immunization program. Not long after the program was launched, cases of vaccine-related Guillain-Barré syndrome began to appear, followed predictably by a spate of lawsuits against the government.
>
> During the same period, the Department of Health, Education and Welfare launched a national initiative for childhood immunizations . . . concurrently, controversy erupted in the professional and public media over the interpretation of available data on the risk of serious injury resulting from the administration of pertussis and other vaccines. Suits against the seven manufacturers then producing vaccines arose with increasing frequency, as did the costs of litigation and settlements, and ultimately the cost of the vaccine itself (Landwirth 1990).

During the early 1980s vaccine injury was the subject of scholarly reports by groups including the U.S. Office of Technology Assessment, the American Medical Association, the American Academy of Pediatrics, the American College of Physicians, and the Institute of Medicine of the National Academy of Sciences.

Because of the perceived risk, the public outcry, and questions about the efficacy of whole-cell pertussis vaccine, the U.S. Congress passed the 1986 National Childhood Vaccine Injury Act (Public Law 99-660, Title III). These developments are sketched from a legal viewpoint in the review by Landwirth (1990), who stated that "the overriding purpose of the National Childhood Vaccine Injury Act is to provide a fair and expeditious legal remedy for children who suffer serious sequelae from mandated vaccinations, to offer greater protection from liability for vaccine administrators and manufacturers, and to ensure a safe and secure vaccine supply."

Among other things, this law mandated the Institute of Medicine of the U.S. National Academy of Sciences to convene a special "Committee to Review the Adverse Consequences of Pertussis and Rubella Vaccines." The Committee evaluated evidence from animal and human experiments, case-comparison, cohort, and other controlled studies, and reports of individual cases and series of cases. The report of this Committee was published as a 367-page book (Howson et al. 1991).

The Act also established a National Vaccine Program, a National Vaccine Advisory Committee, and an Advisory Committee on Childhood Vaccines, and defined a rigorous procedure under which a person could bring a civil action for damages against a vaccine manufacturer. The National Childhood Injury Compensation Program was established effective October 1988. Under this act, persons could be compensated for vaccine-related damages through a National Vaccine Injury Compensation Trust Fund, built up from an excise tax on each unit of vaccine sold. To reflect currently accepted views about the relative reactogenicity of different vaccines, the tax rate ranged from two cents per unit of adult diptheria-tetanus toxoids to $1.56 per dose of pediatric DPT, primarily because of the whole-cell pertussis component. Lederle laboratories increased the price per dose of injectable (killed) polio vaccine from $0.39 to $1.56 between 1981 and 1986. The vaccine cost alone for a complete series of childhood immunizations rose from $4.35 in 1978 (DTP, polio, and MMR) to $112.63 in March, 1992 (including *Haemophilus influenzae* type b which was added in the intervening years and the proposed addition of hepatitis B). Administration costs are not included (Immunization Update, 1992).

A comprehensive law passed by the parliament of Denmark decreed that an individual is entitled to economic compensation for injuries related to vaccination (Von Magnus 1973). To be awarded compensation, an individual must show a "reasonable probability" that a vaccine caused the injury, but demonstration of an absolute causal connection is not required. Before the National Childhood Vaccine Injury Act of 1986, the only avenues of compensation in the United States were through the tort system or private settlements with manufacturers. Like the Danish law, the U.S. act provides compensation on a no-fault basis, as no fault on the part of the manufacturer need be demonstrated. All that is required is that the injured person sustain an injury for which compensation is authorized. Awards cover medical, rehabilitative, and special education expenses, and legal fees for both successful and unsuccessful applicants are to be paid by the program. The law limits awards that could be acquired through the tort system, should individuals reject the compensation offered under the program. Compensation for unavoidable side effects is prohibited under the law, and limits are placed on the award of punitive damages. Drug companies are not presumed to be liable if they have complied with FDA requirements.

Igelhart (1987) observed that the enactment of the compensation program "represented a congressional determination that society, through its government, rather than the vaccine recipients and manufacturers, should assume the responsibility for the administration and funding of this type of compensation."

## Getting Value for Money from Biomedical Innovations

The motivations for investing in the biomedical realm differ for various parties. In the case of commercial organizations there is the expectation of profit. For governments and official health agencies there may be an expectation of better health leading to increased longevity, productivity, and earnings, and averted future costs for medical care. Some investments are highly profitable. The expenditure of $46 million over a period of 12 years to eradicate smallpox resulted in a saving to the United States of about $500 million per year forever (Bloom 1986). Some other authorities place the savings at double that amount. Additional returns on expenditures can come in the general form of good will or favorable opinions, such as political support.

### *Cost-Benefit Analysis (CBA)*

The very large literature on this subject contains many sophisticated methodologies, all of which are directed to the goal of enlightened, or at least informed, decision making in the face of uncertainty. Together with cost-effectiveness analysis (defined later), CBA forces "explicit consideration of the data needed and assumptions made in estimating how resources are, or will be, used to produce the expected results." (Haaga 1983). The meanings of "benefit" have been discussed in chapter 1. In one interpretation, benefits of an innovation accrue most dramatically in the case of life-saving procedures applied to the very young, resulting in the accumulation of many years of productive life gained by the individual involved. Moreover, the prevention of communicable diseases by immunization has external benefits to society by reducing transmission of the pathogen and contributing to the level of herd immunity (chapter 3).

The "complexities and nuances of evaluating the costs associated with providing medical technologies" have been considered at length by Luce and Elixhauser (1990). The costs of fully implementing an innovation, in addition to the obvious direct, indirect, and intangible elements such as salaries, facilities, materials, transportation, time spent by patient's family, psychosocial costs, and so on, must also include a consideration of the possible adverse consequences (risks), an assessment of the likelihood that each will occur, and an estimate of the cost or loss to society whenever that consequence occurs.

*Defining the cost-benefit time horizon.* A common accounting procedure transforms future financial consequences to their net present value. This is the amount of money that needs to be invested today at a specified discount rate (rate of real interest) to total the amount of future benefit that is obtained or forgone.

No general rule exists for selecting the date on which the books are closed in calculating the benefit stream: in the near term the defined benefits may be substantial but the ultimate effects, if any, of preventive procedures are less easily predicted and analyzed. The time horizon that is selected for the analysis can make a crucial difference in calculating the magnitude of costs and benefits.

The safety and efficacy of vaccines are generally evaluated on a relatively short time horizon: immunize, challenge, and conclude within months or a very few years. The need to reach conclusions within a reasonable time is understandable for all the obvious reasons, but some consequences may be unforeseeable. The example has been given of polio vaccine naturally contaminated with SV40 that could have required years to show an adverse effect.

What should be the attitude of the biotechnologist toward the possibility of undefined and unpredictable risks that may extend far into the future? Cavalier dismissal is imprudent, and excessive preoccupation will lead to paralysis and may scuttle a beneficial program. On a practical level it seems reasonable to be aware and vigilant, to discuss frankly with ministries, funding agencies, and others involved the ultimate responsibilities of each party, and to plan for a rational follow-up and surveillance system.

*Costs and consequences of health improvement.* Benefits in terms of lives saved may be foreseen also as costs involved with the population growth resulting from increased survival. It is often pointed out that adding lives to a community or extending the economically productive years has little realizable benefit if there is no employment or other opportunity to be economically productive. Countries unable to provide adequate social services to their current populations may not welcome further additions to their already underemployed workforce. Efforts to provide education, housing, water, and health facilities may be undermined by relentless population increases brought about, at least in part, by biomedical innovations. The environmental degradation resulting from enlarged populations was discussed in chapter 3. There is no need for a moratorium on innovation just because the consequences of health improvement are equivocal. Nevertheless, the introduction of "weapons of mass survival" to countries already overcrowded should be accompanied by enlightened policies to cope with the effects of resultant increases in population.

## Innovations for Prevention

It is usually considered axiomatic that prevention is cheaper than cure. The assertion that immunization is the most cost-effective of all health interventions is repeated like a mantra in innumerable publications, but rarely supported by hard evidence. The cost savings that accrue from preventive activities should really be reviewed for each particular circumstance. In a study titled "Is Prevention Better than Cure?" Russell (1986) stated:

> The costs of a preventive program depend on the size of the population at risk, while its health benefits derive from the smaller number who, in its absence, would actually have contracted the disease. By contrast, only those with the disease incur the costs of acute care and they also derive its health benefits. As a result, the cost per person of acute care can be much higher than those of prevention and still produce the same, or even lower, costs per life saved, or per year of life saved.
>
> The frequency with which a preventive measure must be repeated has a similar effect on costs and benefits. For example, screening tests for high blood pressure, cancer, or other diseases must be repeated at regular intervals if they are to detect disease early enough to be helpful. Each repetition must be applied to the entire population at risk, only a few of whom will be found to have the disease at a given screening.

Assuming that costs and benefits can be measured, or at least estimated, reasonable people may ask what return they may expect from their investment. While it is generally fairly easy to count today's costs, it is at best very difficult to project future health benefits, monetary savings, and increased productivity. If it is certain that a current expenditure will avert all future costs, as in the example of smallpox eradication, everyone would agree that is money well spent.

*Immunization programs.* Many studies have demonstrated the positive cost-benefit ratio of mass immunization programs against common childhood diseases. Given estimates of expected attack rates, probabilities of occurrence of serious manifestations, cost of resultant medical care, and similar information for unprotected populations, credible cost-benefit analyses can be made. For example, there were about 4 million annual cases of childhood measles in the United States in the 1950s, 4,000 cases of measles encephalitis, 1;000 or more cases of permanent mental retardation, and 400 deaths. In the first 5 years after its introduction in 1963, measles vaccine was credited with preventing 9.7 million cases of measles at a saving of 32 million schooldays, 1.6 million workdays, and 555,000 hospital days. An estimated 973 lives were saved, 3,244 cases of mental retardation prevented, and more than $423 million in costs were averted (Axnick et al. 1969). The cost-benefit analysis by Jönsson et al. (1991) of hepatitis B vaccination in Spain was based on a detailed model of the epidemiology of that infection, the cost and efficacy of immunization, expected compliance, anticipated adverse effects, costs of screening, and direct and indirect costs of the illness including quality of life. It was concluded that

> a vaccination program will reduce direct expenditures for hepatitis B if the attack rate in the target population is higher than 4.9%. If indirect costs are included, the threshold for cost saving is reduced to 0.9% (Jönsson et al. 1991).

The Expanded Programme on Immunization (EPI) of the WHO has conducted economic studies of immunization programs in many countries. For example, in

Indonesia, Barnum et al. (1980) analyzed a 5-year campaign to immunize pregnant women against tetanus, and infants against tuberculosis, diphtheria, pertussis, and tetanus. It was concluded that benefits would continue for many years after the formal termination of the campaign, with a total of about 3.5 million cases and 250,000 deaths averted. More directly, program costs of around 13 billion rupiahs would generate a benefit of about 59 billion rupiahs, plus reduced transmission of the diseases, better schoolwork by immunized children, and reduced maternal mortality. Despite these studies, a measurable economic loss cannot always be demonstrated to arise from the presence of a disease, in which case a costly control campaign must be thoroughly justified.

Costs are often difficult to estimate, and national figures for cost per fully immunized child may have little real meaning. In Brazil, Dominguez-Ugá (1988) showed that the average total (fixed plus variable) cost per immunization in six districts of Para state ranged from 378 to 1,068 cruzeiros, and among 11 districts of Pernambuco state it varied from 216 to 1,245 cruzeiros, because of population dispersal and related factors. The cost of the vaccines used in these two states was about 4% of the total immunization costs. In other immunization programs studied, the cost of the vaccine is usually about 10% to 12% of the cost of the entire program (Table 5–4). In Kenya in 1978, labor accounted for 62% of total program costs, and vaccines were the second largest expenditure, with 21% of total costs (Table 5–11). In Latin America, recurrent costs associated with the cold chain can account for as much as 50% of the total costs of immunization programs. Furthermore,

TABLE 5–11.   Cost Analysis of the Expanded Programme
on Immunization, Kenya, 1987

| Item | Cost, $US | Percent of Cost |
|------|-----------|-----------------|
| Annualized capital costs | | |
| Cold chain | 165,784 | 2 |
| Transportation | 173,784 | 3 |
| Immunization equipment | 12,872 | 0 |
| Buildings | 219,734 | 3 |
| Other | 92,045 | 1 |
| Start-up labor cost | 83,401 | 1 |
| Total | 746,964 | 11 |
| Recurrent costs | | |
| Labor | 4,164,147 | 62 |
| Immunization equipment | 136,846 | 2 |
| Vaccines | 1,424,610 | 21 |
| Transportation | 214,116 | 3 |
| Other | 81,210 | 1 |
| Total | 6,020,929 | 89 |
| Total annual cost | 6,767,893 | 100 |

*Source:* Bjerregaard 1991.

Since many vaccines currently available require a syringe for their effective administration, the health care system must either develop or augment waste management procedures for the proper disposing of needles and syringes, set up and develop an infrastructure to train, equip and resupply all health establishments with the sterilization equipment and fuel. The recurrent costs associated with injection technology and logistical supply may be financial and management obstacles and hence impede and disrupt the vaccination program, thereby delaying the achievement of targets (DeQuadros et al. 1992).

A useful summary of immunization problems in developing countries is shown in Table 5–12.

*Cost-effectiveness analysis.* Cost-benefit analysis and maximization of monetary return are insufficient bases for health investment decisions. Public health goals must also be considered, and planners need to find the best means to achieve them. Cost-effectiveness analysis begins with a predefined goal and then works to discover the best way to reach that particular objective: "In CEA alternative courses of action that have similar health benefits are compared, thus reducing the problem primarily to a comparison of costs" (Fuchs 1980). For example, Table 5–13 shows an analysis of the cost-effectiveness of preventing a death from hepatitis by incorporating hepatitis B vaccine into an existing EPI program.

In contrast to the large population-based economic studies is a small-scale analysis of costs and benefits of preexposure rabies prophylaxis. The Peace Corps has experienced about six rabies exposures per year per endemic country, but only

TABLE 5–12.  Immunization Problems in Developing Countries

1. Inadequate surveillance of infectious diseases
2. Inadequate diagnostic facilities
3. Inadequate and unreliable transportation, maintenance problems in the tropical environment, and the difficulties of movement in the rainy season
4. Inadequate health personnel for surveillance, diagnosis, and the delivery of vaccines
5. Inadequate funds for immunization programs
6. Remote and dispersed populations in many areas
7. Problems in record keeping because of illiteracy rate
8. Problems in communication
9. Problems in maintaining the cold chain for vaccines and lack of sufficiently heat-stable preparations
10. Poor antibody response to some vaccines such as OPV because of poor nutrition, poor immune response, presence of inhibitors, interference by other agents, loss of antigenicity in tropical areas, inadequate dose because of faulty equipment, and other unknown reasons
11. Early age of infection, requiring immunization in the first year of life, perhaps earlier, even at the time of birth
12. Difficulty in getting people back for follow-up doses after the first one
13. Poor integration of immunization programs into other health activities
14. Lack of political will and support.
15. Higher priority given to other health or economic programs

*Source:* Evans 1989.

TABLE 5–13.   Cost-Effectiveness of Adding Hepatitis B Vaccine to Current
EPI Antigens

|  | Minimum | Maximum |
|---|---|---|
| Vaccine Cost per 3 doses | $1.50 | $3.00 |
| Delivery cost per antigen | $0.75 | $2.25 |
| Delivery cost plus hepatitis B vaccine | $2.25 | $5.25 |
| Cost to immunize 1,000 children | $2,250.00 | $5,250.00 |
| Number of hepatitis B-related deaths prevented | 30 | 15 |
| Cost per death prevented | $75.00 | $350.00 |

Source: Maynard et al. 1989.

Assumptions: (1) 3 doses of hepatitis B vaccine are provided to 1,000 children for $1.50 to $3 per child for vaccine plus one-sixth of EPI program costs; (2) between 15 and 30 deaths are prevented in this cohort. Costs are in US$.

two deaths from this cause have occurred among 130,000 volunteers since 1962. The cost of a 3-dose series of human diploid cell rabies vaccine is $208.50 in the United States. Considering the exposure risk, the life-threatening nature of rabies, and availability (and cost) of postexposure treatment, is the vaccine cost-effective?

> The use of pre-exposure prophylaxis in travellers was not cost-effective and will not become so until the price of a dose of vaccine declines substantially to $7.00 for the Peace Corps, and even lower for groups with less rabies exposure. However, despite the high vaccine cost, pre-exposure prophylaxis continues to be recommended in the Peace Corps for important non-economic reasons which may also be applicable to other groups of travellers (Bernard and Fishbein 1991).

## Innovations for Diagnosis

Where early diagnosis is feasible, subsequent expensive clinical illness may be avoided. The cost of diagnosing and treating an early case is compared to the costs incurred if the illness proceeds to an advanced stage. The same general principles apply in community-based mass screening programs (for tuberculosis, yaws, malaria, trachoma, esophageal cancer) and in diagnostic procedures on individuals. Additional benefits of early diagnosis are that infections are less likely to be transmitted, and that useful epidemiological information is obtained.

*Increasing costs from more diagnosis.*  The introduction of diagnostic hardware such as CAT scanners often results in a proliferation of tests for suspected brain tumors and similar indications. Some of these tests may be generated by their mere availability, professional curiosity, the practice of defensive medicine, and cost reimbursement policies in various countries. The sudden increase in reporting of a disease, resulting from a diagnostic innovation or availability of technology, must be distinguished from a genuine epidemic.

A new diagnostic method can generate significant second-order expenditures for costs incidental to the procedure itself (professional time, patient travel, waiting time and loss of production, anxiety, taking inappropriate remedies), follow-up or confirmatory tests for inconclusive results or false positives, and the ethical imperative for treatment of those found to be positive. An important element in introduction of a diagnostic innovation is whether the medical care system is capable of providing the patient with an improved outcome.

The question of the most appropriate level of resources to devote to a particular objective has occupied armies of economists and planners. According to one health economist.

> For the health professional, the "optimum" level is the highest level technically attainable, regardless of the cost of reaching it. The economist is preoccupied with the social optimum, however, which he defines as the point at which the value of an additional increment of health exactly equals the cost of the resources required to obtain that increment. For instance, the first few days of hospital stay after major surgery might be extremely valuable for preventing complications and assisting recovery, but at some point the value of each additional day decreases. As soon as the value of an additional day falls below the cost of that day's care, according to the concept of social optimum, the patient should be discharged (Fuchs 1974).

## Implementation

### *Strategy for Routine Use of Immunizations*

All of the points listed in Table 5–12 need to be addressed on a daily basis in routine immunization programs, with emphasis on continuity and sustainability of programs. Ofosu-Amaah (1983) emphasized that African countries were still seeking the most effective and accepted methods for using and administering vaccines. Some countries use "vertical" campaigns with centralized control; others incorporate efforts into maternal and child health programs, or combine mobile health teams with more centralized efforts. Achieving immunization goals with community acceptance can be done by working through the societal infrastructure. In some areas, communication about an immunization program is best done door-to-door; elsewhere, community meetings, village leaders, and school teachers may be more effective (Heggenhougen and Clements 1987). In all areas, managerial capacity, staff training and encouragement, and political commitment remain crucial.

In many countries immunization is both compulsory and free of charge. In Egypt, Imam (1985) noted the relationship between primary health care and immunization, saying that the immunization program encourages the practice of primary health care and that primary health coverage translates into better vaccination coverage. The production and delivery of safe and potent vaccines is strengthened by a national laboratory for the control and production of vaccines, which also has responsibility for other biotechnology.

The greatest reduction in disease incidence will occur when there is a high proportion of susceptibles and substantial risk of exposure and illness, plus:

- Effective health services with an adequate number of trained and motivated personnel and properly distributed facilities
- Affordable and accessible vaccines, syringes, and supplies
- Transportation for staff and the target populations
- Immunization of children when they appear for curative reasons
- Use of mobile immunization teams where needed
- Provision of services at a convenient time and place
- Reduction of the sociocultural differences between project staff and the community by incorporation of known and trusted persons
- Adequate record-keeping and information systems
- Communication of the goals of the program in a comprehensible, nonpatronizing manner
- Administration of several vaccines at the same time

Abundant experience over the past 15 years in every part of the world has stimulated the EPI to reflect on the lessons that it has learned, as a guide to immunization programs in the 1990s. Some of these lessons apply solely to this unique and herculean effort, but others, (see Table 5–14) are more generalizable and may provide valuable insights for those interested in technology transfer and implementation.

Working with a limited amount of meningitis vaccine in northern Nigeria,

TABLE 5–14.    Some Lessions Learned by the EPI to Guide Immunization Programs
in the 1990s

- Nearly all children of the world and their mothers can be reached with immunization services which provide an excellent entry point for other primary health care interventions and significantly reduce mortality, morbidity, and disability.
- Personal involvement of heads of state and political, religious, and social leadership at all levels generates political will, creates demand for immunization services, mobilizes communities, and can even overcome such obstacles to immunization as low levels of socioeconomic development and civil war (for instance, promoting "days of tranquility").
- Communication is a critical component of immunization programs and should be planned and designed as an integral part of the program. Successful programs are based on audience research and use multiple channels (face-to-face, print, mass media) to deliver the action-oriented message.
- Inclusion of only essential and proven cost-effective elements in immunization interventions is a factor in their success.
- Successful programs have a process measure (e.g., immunization coverage) as well as an outcome indicator (e.g., disease incidence).
- Decentralization of resources, coupled with national plans of action, permit countries to plan and execute activities more effectively and rapidly.
- Research and development in logistics, the cold chain, injection equipment, new and improved vaccines, delivery strategies, immunization schedules, and monitoring and evaluation methodologies provide a technical basis for advancing program policies and strategies.
- Training provides the technical skills for planning and management to strengthen the health infrastructure and to develop the needed human resources at senior, middle, and peripheral levels.
- Emphasis on surveillance strengthens all aspects of an immunization program and helps to focus efforts on areas at highest risk.
- Unwarranted contraindications to immunization, unbalanced concerns over adverse reactions of vaccines compared to the risks of disease, and public apathy toward immunization as disease incidence declines are phenomena, seen especially in developed countries, that must be countered everywhere.

Source: Adapted from Expanded Programme on Immunization 1992b.

Greenwood and Wali (1980) vaccinated children in several villages after two cases had been reported. Selective immunization appeared to control the spread of meningitis relative to unvaccinated control villages. Noting that meningococcal disease was most common in 5 year olds, the authors surmised that vaccination of this group only would probably control disease transmission in nonepidemic periods. Selective vaccination by incidence (i.e., waiting until a village reports a case and then responding) demands both effective surveillance and an immediate supply of vaccine.

Much depends on local conditions and social customs. In prewar Kuwait, large-scale rubella vaccinations were probably unnecessary because immunity was high and only 5.4% of 1054 tested women of childbearing age were seronegative for rubella antibodies (Hathout et al. 1978). In light of the low rate of premarital pregnancy in that country, testing for antibodies may best be done just before marriage, and only susceptible seronegative women immunized. Such a strategy would not be useful in other cultures, such as in North America.

*Avoiding vaccine wastage.* Everyone agrees that vaccines are precious and must be handled with great care. Because it is impossible, without sophisticated laboratory studies, to tell whether the vaccine in a particular vial is still potent, arbitrary guidelines are often used to determine when to discard vaccine. The many different vaccines with varying characteristics (Table 5–7) cause confusion among professional and field staff alike. Details vary from country to country, but criteria for discard usually include vials that are expired, frozen, opened, or exposed to heat.

The first case is the simplest, as an expiration date on the label is unambiguous. "Pull dates" are generally very conservative, and although properly stored vaccine may retain its potency well beyond its nominal expiration, no case will be made for the use of outdated vaccines. The best way to avoid vaccine wastage for this reason is proper inventory control and ordering just the right amount of vaccine. In many developing country situations this is much easier said than done.

Vaccines vary in their tolerance to freezing. Oral polio vaccine may be stored frozen (below −20°C) for up to two years and at refrigerator temperatures for 6 to 12 months. BCG can also be frozen. On the other hand, adsorbed toxoids and adjuvanted vaccines are adversely affected by freezing (Galazka 1989). Staff training with frequent reinforcement appears to be the best way to manage vaccine wastage from inappropriate freezing. Another important factor is checking and maintaining equipment to assure that refrigerators and freezers are operating at the proper temperature.

In the case of opened vials, the hazard is contamination rather than deterioration. Even where the rubber seal is in place, microorganisms may infiltrate through various practices of the immunization staff that breach the rules of strict sterile technique. This is particularly true of vaccines that are reconstituted with diluent and those that lack preservatives. Disposing of used vials at the end of each session, or each half day, is reasonable and prudent. The best way to minimize wastage for this reason is to order smaller vials and use them quickly. However, in considering total vaccine costs, some wastage may be tolerable. A careful calculation of the price of 1, 5, and 10 dose vials may show that even with 60% wastage, on average, buying vaccine in 10-dose vials is most cost effective.

Heat exposure of vaccines has been the most difficult issue on which to issue clear instructions. EPI programs have always placed great emphasis on the cold chain, and most workers are very diligent to maintain vaccines at low temperatures. In the absence of an objective indicator of heat exposure, immunization staff are left with arbitrary rules such as, "throw away any vial removed three times from the refrigerator," or "throw away any vial that has been taken from the clinic to the field." These unduly conservative rules have resulted in the wastage of large volumes of perfectly good vaccines. Until fully thermostable polio and measles vaccines are available, the best solution appears to be the addition to the label of indicator dots that show a clear color change in response to cumulative time and temperature exposure. The main problem then is to convince the program directors that it would be cost effective to (1) spend more to purchase vaccines with the heat exposure indicator, and (2) rewrite their training manuals and retrain the field staff to take advantage of this new technology; and to convince the workers to rely on the indicator color rather than on the rules that they had previously been taught.

## Mass Campaigns

Immunization projects should target the susceptible population not just when there is an epidemic, but to confer a continuing measure of protection even on those not vaccinated through herd immunity and reduced prevalence in the population. The most pressing priority in mass campaigns is increasing coverage (Henderson 1984b). Reducing the incidence of a disease tends to raise the age of the first infection, allowing more latitude in administration and timing. Immunization schedules ideally should entail few doses and as early administration as possible, if there is no interference from maternal antibody.

*Advantages and disadvantages of mass campaigns.* Sabin (1980) pointed out that the simultaneous administration of OPV to as many children as possible in the shortest time interval induces the greatest herd immunity as well as individual protection for those vaccinated. This blitzkrieg immunization results in a more extensive excretion of attenuated vaccine polioviruses than even the dissemination of excreted wild type viruses during an epidemic. Consequently, the virulent viruses might be displaced by the vaccine virus. The administration of OPV on a continuing year-round basis fails to provide this "flushing" effect and may not be as cost-effective. The ability of the mass campaign to break the chain of virus transmission appears contingent on a coverage rate of 70% immunization of susceptibles. In areas with only 50% coverage, protection of unvaccinated individuals was not induced.

As mentioned earlier, OPV may fail to stimulate a high antipolio antibody titer in many children in tropical areas. This failure has been ascribed to competition from other enteroviruses, or possibly to overstimulation of the immune system in children in poorly sanitated environments. Further studies are needed to clarify the specific causes of this phenomenon.

Although wide administration of a vaccine within a short period provides a basis for herd immunity, immunization of children born after a mass campaign must be continued on a regular basis if the protection is to be maintained. Failure to vaccinate infants born subsequent to the campaign, as well as the migration of susceptibles, particularly from areas of high poliovirus prevalence, may contribute to a high incidence of disease between mass campaigns.

The induction of herd immunity has the further advantage of protecting those children who, though vaccinated, may have had an enteric infection that interfered with the effectiveness of the vaccine. Administration of vaccine to all children in the most susceptible age group, regardless of immunization history, enhances coverage of unvaccinated children in addition to reinforcing the intestinal resistance of those already vaccinated previously. The target age group is best determined by ascertaining the age at onset of illness of 90% of the cases reported during the previous five years, presuming the availability of reliable records, or verifying the age at which 90% of paralysis victims incurred lameness. Sabin's suggestion that all children receive the vaccine twice a year regardless of vaccination status eliminates the need to maintain records of vaccination status and allows the planners of campaigns more latitude in choosing the starting age for vaccination.

On the other hand, some authorities have criticized mass campaigns. According to Hall et al. (1990), "The measles campaigns in West Africa had shown clearly that, although the disease could be controlled in the short term by mass campaigns, the gains were not sustained and a continuous service was necessary." In South Africa, Barron et al. (1987), while acknowledging the "fire-fighting" aspects of mass campaigns, criticize their lack of sustainability and lack of integration into existing health services. The diversion of workers from other areas of the health service "can lead to duplication of activity and inefficient use of resources since each programme needs its own infrastructure and administration."

*Follow-up of mass campaigns.* After a mass measles immunization campaign in rural Ghana, Belcher et al. (1978) obtained comments from mothers. They found that participation could be improved by attention to the following points:

1. Clearer articulation of the program
2. Specification of the expected benefits—which diseases can actually be prevented
3. The criteria for vaccination (several mothers thought their children were too young to be vaccinated)
4. Longer stays in one place by the vaccination team
5. Use of a vaccine with fewer adverse effects, as more than half of the mothers who declined to attend a follow-up campaign cited the child's pain and swelling as the primary reason
6. Follow-up care by the team to treat symptomatic vaccinees

Belcher et al. (1978) recommended the incorporation of village volunteers and community leaders into publicity campaigns, reiteration of publicity one day before team arrival, and emphasis on diseases familiar to the indigenous population.

A mass measles immunization campaign conducted in 1971 in Yaounde, Cameroon was evaluated by McBean et al. (1976). They found the usual difficulties: lack of attendance, errors in selection of subjects, and only 40% seroconversion because of ineffective vaccine. By their calculations (Table 5–15) 83% of the vaccine was wasted. However, although 44% of vaccinees were found to be already immune because of previous natural infection, the administration of an extra dose to such children is hardly a major sin and could have a beneficial booster effect. The prior determination of seronegativity as a condition of immunization is infeasible and would cost many times more than the "wasted" vaccine given to an already immune child. In the research described, a drop of blood from each child was captured on filter paper, which was then sent to the United States for antibody determination. While it is clearly impossible to conduct such elaborate studies on a routine basis, a strong case can be made for awareness by local authorities of the immune status of the vaccinee population, together with some surveillance of seroconversion rates.

The obvious way to increase the efficacy of mass campaigns is through improvements in screening, handling of the vaccine, and health education of the public.

In Chile, an advanced developing country, Borgoño and Corey reported as long ago as 1978 the success of the polio vaccination campaign. The combination of an active role of the community, a permanent and consistent program, constant epidemiological surveillance, and consideration of immunization as a health priority at the local level helped to reduce the incidence of polio to zero. The successful program overcame previous low interest in the community and a weak health infrastructure that undermined the campaign until the early 1970s, so that individuals were still contracting polio. Immunization of newborns was initiated and was complemented with campaigns for children under 6. Increased urbanization in Chile between 1950 and 1977 also concentrated the population and increased access to the pool of infants.

*Emergency applications.* The difficult conditions intrinsic to natural disasters, wartime hostilities, and refugee camps require flexibility and ingenuity in providing all health services. The risk of certain communicable diseases may be greatly increased because of the disruption in sanitation and living conditions, requiring

TABLE 5–15.   Effectiveness of Utilization of Measles Vaccine During an Immunization Program, Republic of Cameroon, 1970–1971

| Result of Utilization | Percent |
|---|---|
| Given to wrong aged children, or discarded unused after dilution | 14 |
| Given to children who already had antibody | 44 |
| Failed to induce antibody titer | 25 |
| Achieved successful seroconversion | 17 |

*Source:* Adapted from McBean et al. 1976.

protection of health workers as well as of the public (Steffen et al. 1991). The number of refugee and internally displaced people throughout the world is so great that Dyke (1991) has referred to the new medical specialty of "migration medicine." Mention should be made also of the risk of introduction and spread of the agents of tuberculosis, hepatitis B, and perhaps other diseases from refugees repatriated in industrialized countries, and the corresponding need for epidemiologic surveillance.

## Underachievement of Immunization Programs in Developing Countries

Immunization programs in developing countries are often presumed to be effective, but ascertainment of the degree of success is not easy. In the great majority of both routine programs and mass campaigns the true accomplishments are unknown. Where careful studies have been done, it is usually conceded that programs have been only partially successful. It is relatively straightforward to measure the consumption of resources and the production of services, such as the number of immunizations given. Longer-range goals, in terms of health improvement, are often not spelled out, so that the extent to which program goals have been attained cannot be evaluated. The objective may be a general reduction in infant and childhood mortality rates, or in incidence of the particular diseases immunized against. In either case an appropriate surveillance system must be put in place, a system capable of detection and correct identification of the causes of illnesses and deaths in the community. Even if resources were available for such monitoring, which is unlikely, it is not obvious how long it should continue.

Other problems with immunization programs include administration of inadequate doses in error or in order to try to make the vaccine go further, poor planning, lack of attendance, and operational errors. The effectiveness of many programs is marginal because vaccines are given to people who will not benefit from them, for example, those too young or too old.

*Immunization and mortality.* The Expanded Programme on Immunization has set high goals for coverage with its six antigens. In some developing tropical countries programs have been very successful. Greenwood et al. (1987) reported that the proportion of fully immunized infants in The Gambia was higher than in the United Kingdom (see also chapter 1). Although a minority of infant and young child deaths in that country occur in a hospital, these authors attempted, through a house-to-house survey, to interview the family of each child who had died in a rural area near Farafenni to establish the cause of death as precisely as possible. In that region 90% of children aged 3 months to 4 years had a health card, and a high proportion were fully immunized.

Over the span of 1 year 184 deaths of children under age 7 were recorded in this rural Gambian population. As far as could be determined, immunization rates were not statistically different between children who had died and those who had not: only 8 children died from a condition that might have been prevented by routine infant immunization.

It seems likely that the establishment of an effective EPI programme in the Farafeni area has had only a modest effect on infant mortality (142 per 1000) but a more marked effect on CMR (43 per 1000 per year) which is now considerably lower than the levels recorded in these earlier studies in The Gambia and Senegal. Measles vaccination has probably played an important part in bringing about this improvement. . . . A recent review suggests that the Gambian EPI has been very cost effective in preventing deaths from measles. . . . Our data suggest that further improvements in vaccine coverage are likely to have only a modest effect on infant and childhood mortality in The Gambia and that if mortality is to be lowered it will be necessary to concentrate on malaria, acute respiratory infections, chronic diarrhea and deaths in the neonatal period (Greenwood et al. 1987).

## Integration of the Innovation into Existing Health Services

The first problem in introducing an innovation into primary health care programs (see chapter 2) is discovering whether the primary health care programs themselves are well developed and integrated into existing health services. That this is not always the case was emphasized by Fendall (1987): "Despite widespread acceptance of the concept of Primary Health Care (PHC), in practice it has yet to achieve a sufficiently extensive impact in developing countries. . . . The present dual approach of a general health service and a semi-autonomous system of vertical programmes for specific objectives presents problems in developing a coordinated or integrated organisation and management infrastructure."

An outsider, even with a demonstrably beneficial new product, cannot propose to overhaul an existing medical care system. Indeed, it may be easier to change the physical environment than to convince governmental administrators to reorganize their departments. A plethora of separate programs with their own administrations and overlapping and competing jurisdictions, while less than ideal, may present both obstacles and opportunities to the truly innovative innovator.

Circumstances vary so widely that it is impossible to generalize, but various scenarios can be visualized provided that external funding is available. Much depends on the extent of development of the existing health infrastructure, the acceptability of the innovation to the target population, the likelihood of sustainability of the innovation, and similar factors. Existing fixed costs of buildings, equipment, vehicles, and so forth may be shared to accommodate trials or implementation of a newly transferred innovation. Staff from stagnant or terminating programs could be retrained, and public participation can be stimulated by provision of additional benefits.

# 6

# The Global Vaccine Establishment

In contrast to pharmaceutical drugs, which depend on a huge repeat market, a vaccine is used just once, or perhaps a few times, over the human life span. On the other hand, every person should be immunized whereas relatively few will take any one drug. Nevertheless, many vaccines, such as those for plague or rabies, are administered rarely and selectively, on an individual rather than a mass basis. The so-called orphan vaccines against Q fever, Japanese B encephalitis, and the like, although safe and efficacious, have virtually no market in most countries.

Despite the successful introduction of inactivated poliomyelitis vaccine in 1954, the appearance of live polio and measles vaccines in the succeeding decade, and the many more recent technical advances, the interest of pharmaceutical companies in the manufacture of vaccines declined steadily through the 1970s and middle 1980s.

By the early to mid-1980s the gloomy profit picture, exposure to legal liability, organized anti-immunization efforts, shortages of supply owing to technical difficulties in production, and the vagaries of government regulation led to abandonment of the field by one manufacturer after another. The resultant severe vaccine crisis in the United States led directly to the National Childhood Vaccine Injury Act of 1986, discussed in the previous chapter. In many other industrialized countries the situation was no more encouraging, and it became clear to many observers that concerted action was urgently needed.

## The International Framework for Vaccine Research and Management

### Current Leadership in the World Vaccine Picture

The locus of decision making in vaccines was formerly diffused among researchers acting independently, manufacturers who decided what to make, regulatory agencies that acted with fingers crossed under clouds of ignorance, and in some countries unsupervised government laboratories that made vaccines of variable and often unacceptable quality for their official use.

From the late 1980s, research and development on vaccines has been revived in part because of the recognition that new biomedical technologies might improve existing vaccines and create novel products to protect against currently unvaccinable diseases. Equally important is the extraordinary blossoming of national, regional, and global organizations for coordination toward the common goal of universal childhood immunization (UCI). In general, all of these new groups can claim descent from an interlocking set of multilateral and national development assistance agencies, private foundations, voluntary organizations, academic institutions, and industrial companies.

The pervasive characteristic of these groups is that vaccine research and implementation are no longer uncoordinated separate activities, but are deliberately planned and integrated. Needs are identified and defined in advance. Efforts are directed specifically toward predetermined ends, either disease-based (meningococcal disease, malaria) or logistic-based (simpler schedules, fewer doses, longer shelf life, less reactogenic products, sustainable and affordable delivery systems), or both. Much reliance is placed on technology to overcome universally recognized shortcomings in the current world vaccine picture. In addition, more subtle underlying themes among the members of the various panels are a strong belief in the value and cost-effectiveness of vaccines, and a vision of social equity that considers that access should not be denied because of (1) the inability of vaccines to compete in the marketplace, or (2) the poverty of populations that would benefit most from their use. Accordingly, vaccines are viewed not only as commercial commodities but, collectively, as a public benefit like education.

The vaccine-related bodies function entirely through the voluntary agreement of their members. Lacking official policing or enforcement power, their authority is derived from the prestige of their sponsors, the professionalism of their leadership, and the enthusiasm of the participants. Some observers have viewed these organizations as patronizing or conspiratorial or worse. They have criticized priorities that are determined by a self-selected group of First World academicians and bureaucrats. On balance, however, it is likely that the world will benefit from these activities.

On a more pragmatic level, the vaccine-related groups must raise funds for their own activities. Many also obtain and allocate money to support directed research on a variety of vaccine-related activities from basic science to field trials to implementation. The economic element is so pervasive that one senior bureaucrat remarked, "The scene is driven by economics, not epidemiology. People are fighting among each other over a finite amount of money for immunization."

*The role of developing-country governments.* While officials of developing countries are customarily granted the chairmanship of meetings, conferences, panels, committees, task forces, and workshops, important decisions are clearly in the hands of the sponsoring organizations. Nevertheless, various parties including producers, intermediaries, and implementing governments are talking with each other in newfound collaboration. Through it all the end users—the public—remain conspicuously absent.

## The United Nations and Its Agencies

The worldwide enterprise in biotechnology includes many private, national, and intergovernmental organizations whose interests cover a wide range of activities. Among these, a number of United Nations agencies have been particularly active (Zilinskas 1987, 1989). Their complex programs coordinate and interact with those of other UN agencies, other intergovernmental organizations, and outside groups of various kinds.

*World Health Organization (WHO).* The WHO has played a major role in immunization programs by encouraging and facilitating collaboration, coordinating the work through secretariat functions, and developing and promoting standards. In late 1993, the Director-General of WHO announced the merger of the Expanded Programme on Immunization and the Programme for Vaccine Development into a new *Special Programme for Vaccines,* intended to strengthen WHO's capabilities related to technology transfer, quality control, and financing. The new Special Programme coordinates WHO's activities in the rapidly expanding Children's Vaccine Initiative (CVI). The historical development and composition of each of these programs is described separately in the following pages.

1. THE EXPANDED PROGRAMME ON IMMUNIZATION (EPI). When the Expanded Programme on Immunization was launched by WHO in 1974, there were few immunization services in most developing countries. The EPI has developed in close collaboration with UNICEF, which provides the vaccines and much of the supplies and equipment needed for their administration. Financial support is obtained also from various United Nations agencies, multi- and bilateral donor organizations, and many UN member governments. In addition, private groups such as the Rockefeller Foundation and Save the Children Fund, and organizations such as Rotary International contribute funds through WHO or UNICEF, or provide logistic or other assistance.

From the beginning, the six "target" diseases included in the EPI program have been diphtheria, pertussis, tetanus (combined in the DPT triple vaccine), measles, polio, and tuberculosis, using BCG vaccine. A Global Advisory Group (GAG) has developed EPI strategies, such as the Plan of Action (1982) for establishing national EPI infrastructures. This plan called for promoting EPI in the context of Primary Health Care (see chapter 2); investing sufficient human and financial resources; ensuring adequate evaluation; and including research as a part of program operations. This plan and subsequent GAG policy recommendations have been endorsed by the member states of the World Health Organization through resolutions passed at the World Health Assemblies held every May in Geneva.

EPI management feels strongly that social equity demands that all people be treated equally, with high coverage everywhere. Priorities are to (1) sustain what has been achieved; (2) achieve disease control for polio, measles, and tetanus; (3)

add new vaccines such as hepatitis B; and (4) make the schedules of contacts for EPI and other vaccines more practical.

The eventual goal of EPI is universal childhood immunization which means 100% coverage, but the EPI established intermediate targets of 80% by 1990 and 90% by 2000. Henderson (1990) estimated that worldwide immunization coverage had reached 71% for BCG at birth, about 67% for a third dose of polio or DPT vaccine, and 61% for measles vaccine during the first year of life. By April, 1992, using its own sources, the EPI estimated much higher coverage rates (Table 6–1). They stated that 1.65 million deaths from measles, and 650,000 deaths from pertussis were being prevented each year. The EPI also estimated that 793,000 deaths from neonatal tetanus were averted, despite the low (39%) global coverage of pregnant women with tetanus toxoid. Nevertheless, about 1.9 million annual deaths may still occur from these three causes alone. Approximately 530,000 cases of polio are said to have been prevented.

Among other concerns, the EPI has made a major effort to improve cold chain technology and thermostability of vaccines intended for administration in remote regions of tropical countries. Proper handling and use of vaccines are standardized and facilitated by training materials produced in collaboration with governments and various organizations. In addition, many technical publications are issued for health professionals, and educational materials for the general public.

Few people believed, when the programme was created, that the 1990 goal of providing immunization for all children of the world was anything but wishful thinking. However, this initiative, like the smallpox eradication programme before it, is providing a compelling demonstration of what can be accomplished when there is unanimity of purpose. . . . This has been possible because the programme is easily understood, inexpensive and easy to implement, and because it brings immediate, highly visible benefits. It is good public health and good politics (Henderson et al. 1988).

TABLE 6–1.   Estimated Percentage of Children Immunized in the First Year of Life and Percentage of Pregnant Women Immunized against Tetanus, by WHO Region, April 1992

| Region | Percentage of Children Immunized by 12 Months of Age | | | | Percentage of Pregnant Women Immunized |
|---|---|---|---|---|---|
| | BCG | DPT3 | Polio3 | Measles | |
| African | 81 | 58 | 58 | 58 | 48 |
| American | 82 | 76 | 89 | 81 | 31 |
| Eastern Mediterranean | 86 | 82 | 81 | 81 | 62 |
| European | 74 | 80 | 82 | 79 | 2 |
| Southeast Asian | 93 | 88 | 90 | 80 | 67 |
| Western Pacific | 96 | 94 | 94 | 93 | 6 |
| Global | 88 | 82 | 84 | 80 | 39 |

Source: EPI documents.

Despite its achievements, the EPI is faced by many near-term challenges, such as:

- Achieving and sustaining universal immunization coverage with all the antigens used by EPI
- Controlling the target diseases, including the global eradication of poliomyelitis by the year 2000, the reduction of measles by 90% compared with preimmunization levels, and the elimination of neonatal tetanus by 1995
- Improving surveillance to provide accurate assessment of the progress of the program
- Incorporating new or improved vaccines as part of routine national immunization services
- Promoting other practices that are appropriate for the primary health care system and local communities
- Continuing research and development, particularly operations research to strengthen delivery systems

2. THE SPECIAL PROGRAMME FOR RESEARCH AND TRAINING IN TROPICAL DISEASES (TDR). This program deals with six diseases: malaria; trypanosomiasis, both African (sleeping sickness) and American (Chagas' disease); leishmaniasis; schistosomiasis; filariasis; and leprosy. Work is supported also in social and economic research and epidemiologic studies, research training, and institution strengthening related to these diseases. Since 1976 the TDR has supported thousands of research projects in academic and research institutes throughout the world. It has also collaborated actively with industry for development and testing of agents that might be useful for diagnosis, treatment, or control of the six diseases. Furthermore, TDR has conducted symposia on the transfer to national health services of technology developed with TDR support.

In 1988 the TDR established an Initiative for Biotechnology Implementation (IBI), to consist of a limited number of collaborative partnerships between institutions in the developed and developing world (Bialy 1988).

3. THE DIVISION OF DIARRHEAL AND ACUTE RESPIRATORY DISEASE CONTROL (CDR). This agency embraces separate staffs of personnel specializing in diarrheal or respiratory diseases, particularly those occurring in early childhood and having viral or bacterial etiology. This combined unit collaborates in ongoing work toward development and evaluation of relevant vaccines such as those against rotavirus diarrhea, and *Haemophilus, Pneumococcus,* and other major respiratory pathogens.

4. THE PROGRAMME FOR VACCINE DEVELOPMENT (PVD). This program, initiated by the WHO Director-General in 1984, has received financial support from the United Nations Development Programme (UNDP) since 1990. Additional funding comes from various governments and private foundations. Objectives of the PVD are:

1. Improve existing vaccines which, because of various shortcomings such as reduced shelf life at tropical temperatures, high cost, and limited efficacy, are not totally effective in the developing world.

2. Develop new vaccines against major viral and bacterial diseases where no such vaccines currently exist.
3. Provide mechanisms for international collaboration and coordinate broad participation of the public and private sectors.
4. Develop general methods to improve all vaccines. This program, called the Transdisease Vaccinology (TDV) component of the PVD, was established in 1987 to deal with general issues common to many vaccines, such as simplified vaccine delivery, reduction of multiple doses to single doses, enhanced immunogenicity of vaccines, and so forth. This program strives to develop encapsulated vaccines for slow and controlled release; to develop systems for oral rather than injectable vaccines; and to evaluate the acceptability and efficacy of genetically engineered live viruses and bacteria as delivery vehicles (Bektimirov et al. 1990).

The PVD's long-term goal is "to minimize the number of inoculations needed, and to ultimately make a 'children's vaccine' containing many immunizing vaccines all delivered in a single dose at a time soon after birth." The work of the PVD is closely related to other research-based programs of the WHO, functioning to a large extent through five steering committees for tuberculosis, acute viral respiratory diseases of childhood and measles, dengue and Japanese B encephalitis, viral hepatitis A and poliomyelitis, and encapsulated bacteria (bacterial meningitis and pneumonia). This program emphasizes recombinant DNA technology, peptide synthesis, and other advanced techniques as well as production of live vaccines through methods of genetic engineering. Emphasis is placed on safety, efficacy, and heat-stability of new vaccines. Strategic plans developed by each steering committee call for the achievement of specific technical goals leading to development and testing of improved vaccines. Major vaccine targets are:

1. Improved vaccines needed because current vaccines are not fully effective in developing countries.
   A. Not fully effective: tuberculosis, cholera, typhoid
   B. Not fully effective in young children: measles, group C meningococcal meningitis
   C. Not fully effective in developing countries: oral polio
   D. Too costly for developing countries: Japanese B encephalitis
2. New vaccines needed because no effective vaccine exists.
   A. Viral diseases: dengue, rotavirus diarrhea, hepatitis A and E, acute respiratory infections
   B. Bacterial diseases: bacterial diarrhea, group B meningococcal infections
3. Improved methods for vaccine development that allow easier delivery and result in more effective immunization in the developing world.
   A. Single-dose or one-shot vaccines
   B. Oral vaccines
   C. Multiple vaccines delivered in one dose or one shot

In 1991–1992 eighty extramural research projects were funded by the PVD in the five Task Force areas and in Transdisease Vaccinology. Table 6–2 shows specific targets of the PVD to be achieved before 1995 and 1999.

Through the PVD, the WHO has established immunology training programs at the WHO Immunology Research and Training Centre (IRTC) in Lausanne, Swit-

TABLE 6–2.  The Programme for Vaccine Development: Vaccine Targets
for 1995 and 1999

| Pathogen | Current Situation | Before 1995 | Before 1999 |
|---|---|---|---|
| *New Bacterial Vaccines* | | | |
| Meningococci A and C | Candidate conjugate vaccines (3 injections) | Single-dose candidate vaccines | Into multivalent vaccine |
| Meningococci B | Candidate vaccines | Single-dose candidate vaccine | Into multivalent vaccine |
| *Haemophilus influenzae* | *H. influenzae* b conjugate | Single-dose vaccine | Into multivalent vaccine |
| *Streptococcus pneumoniae* | Vaccine not protective in infants | Conjugated vaccine | Into multivalent vaccine |
| *Salmonella typhi* | Candidate vaccine | Single-dose candidate vaccine | Into multivalent vaccine |
| *Shigella and cholera* | Research in progress | Candidate vaccine | — |
| *New Viral Vaccines* | | | |
| Respiratory syncyctial virus | Candidate vaccines available | Vaccine available | Single-dose vaccine |
| Rotavirus | Candidate vaccines available | Vaccine available | Single-dose vaccine |
| Hepatitis A | Candidate vaccines | — | Oral attenuated vaccine |
| Hepatitis C and E | Research in progress | Candidate vaccine available | Vaccine available |
| Dengue | Single-type candidate vaccines | Multi-type candidate vaccines | Vaccine available |
| *Improved Vaccines* | | | |
| Tetanus toxoid (for NNT[a]) | 3–5 injections | Encapsulated single-dose | — |
| Hepatitis B | 3 injections | | Incorporation into a single-dose multivalent vaccine |
| Diphtheria/Pertussis/Tetanus | 3 injections | Single-dose DPT | |
| Poliomyelitis | Heat-sensitive; 3 oral doses | Heat-stable; oral | Single-dose oral |
| BCG | Variable efficacy | — | Engineered BCG, more efficacious |
| Measles | Injected at 6–9 months | — | Oral, soon after birth |
| Japanese encephalitis | 3 injections at 12 months | Single-dose candidate vaccine | Single-dose, soon after birth |

*Source:* PVD documents.
[a] Neonatal tetanus

187

zerland. At this site an advanced course titled "Immunology, Vaccinology and Biotechnology Applied to Infectious Diseases" is given twice annually, once in French and once in English. The government of Switzerland funds the program, organizes the courses, and brings trainees for 1 to 6 months to attend the course and obtain other specific training. About 40 sponsored fellows participate each year plus a few who come with their own funds, with a cumulative total of approximately 600 scientists from 86 countries. The program wants to expand training into vaccine evaluation and diagnosis of infectious diseases. The PVD has also sponsored regional conferences in Costa Rica, Gabon, Indonesia, and elsewhere.

Policies are determined by a Scientific Advisory Group of Experts (SAGE) consisting of 16 world experts in vaccine development. They represent health ministries, research institutes, foundations, universities, and clinical facilities. An effort is made to achieve continuous integration with end users of the vaccines and to maintain coordination with the WHO Executive Board, World Health Assembly, and other groups both within and outside the WHO.

Interactions have been established with vaccine manufacturers in three different patterns. In the first, a vaccine is developed by manufacturers with the advice of a PVD task force that organizes phase I and II trials in endemic areas (e.g., meningococcal conjugate vaccine). In the second pattern, a vaccine is developed by the manufacturer on the basis of PVD-promoted research. Collaboration is established for preclinical and clinical development (e.g., respiratory syncytial virus vaccine). Alternatively, a vaccine is developed in initial close collaboration between PVD and the manufacturer from the initial stage (e.g., single-dose tetanus toxoid).

5. THE CHILDREN'S VACCINE INITIATIVE (CVI). Former EPI director D. A. Henderson is said to have conceived the idea of a type of "Manhattan Project" for super polio vaccines to achieve polio eradication. He promoted this concept vigorously in various agencies, including UNICEF, in which the idea mutated to the concept of the "single shot" children's vaccine. The Task Force for Child Survival, based at the Carter Center in Atlanta, Georgia, provided significant impetus through the founding document of the CVI, the Declaration of New York, (see box) which was issued at the World Summit for Children held at the General Assembly of the United Nations in September 1990.

The CVI is funded primarily by UNICEF, the Rockefeller Foundation, WHO, UNDP, and the World Bank. The executing agency is WHO, which provides a small secretariat and office facilities in its Geneva headquarters building. A continuing effort is made to attract funds from other agencies.

The Initiative promotes planning, programming, and monitoring activities that facilitate increased cooperation among potential collaborators in the public and private sectors, and in the vaccine research, development, delivery, and regulation communities. Specifically the CVI aims to support the development of the "Children's Vaccine," a core vaccine in a single dose with up to 14 antigens against major viral and bacterial diseases. The composition of this vaccine would vary by

# DECLARATION OF NEW YORK
# THE CHILDREN'S VACCINE INITIATIVE
## 10 September 1990

Children represent the most vulnerable segment of every society—and they are our present and the future. Good health, especially of children, promotes personal and national development. Scientific progress, matched with improved capacities of all countries to immunize their children, provides an unparalleled opportunity to save additional lives and prevent additional millions of disabilities annually through a global "Children's Vaccine initiative".

Working together, national and international agencies, service organizations and private voluntary agencies around the world have demonstrated the feasibility of providing immunization to over 70% of children in the developing world, saving approximately two to three million children per year from those preventable diseases. Furthermore, countries around the world have demonstrated public interest in mobilizing to protect their children. With universal immunization using the current vaccines, two to three million more deaths could be prevented annually. Development of new vaccines against diseases could save another five to six million lives annually during this decade.

Universal immunization will be facilitated by accelerating the application of current science to make new and better vaccines, benefiting children in all countries. These include vaccines which:

- require one or two rather than multiple doses;
- can be given earlier in life;
- can be combined in novel ways, reducing the number of injections or visits required;
- are more heat stable, retaining potency during transport and storage, particularly at tropical temperatures;
- are effective against a wide variety of diseases not currently targeted by immunization but which takes a heavy toll of needless deaths, including AIDS, acute respiratory infections, diarrhoeas, and important parasites; and
- are affordable.

Work on vaccines themselves must be accompanied by investments which bring them rapidly into large scale and inexpensive production and effective use. Such investments are needed to simplify production and quality control methods; to support field trials; to speed licensing; to develop approaches—including production in developing countries—which assure that vaccines are available for all, to simplify the logistics of storing, transporting and administering vaccine, and to strengthen national epidemiological and applied research capacities, especially in developing countries, so that each vaccine is used to best advantage.

We, an international group of experts in the field of research and application of vaccines, acting in our individual capacities to review the great contribution which new and improved vaccines can make during the current decade to the health of all the world's children;

Urge national leaders, the heads of national and international agencies concerned with vaccine development and use, commercial enterprises, and private and voluntary groups—harnessing the scientific and technical capacity of the world, North and South—to commit themselves to a Children's Vaccine Initiative which aims to produce and deliver "Ideal children's vaccines", which: provide lasting protection against a wide range of diseases, need fewer contacts, have more heat stability, are simpler to administer and are affordable.

And,

Request the World Health Organization, in fulfillment of its constitutional mandate as the directing and coordinating authority on international health work, to catalyze global efforts toward these ends by taking the lead in establishing an International Task Force for Vaccine Development with UNDP, UNICEF, The World Bank, and other interested international, national, public and private organizations to coordinate their efforts through exchanging information, agreeing upon priorities, monitoring and evaluating research results, helping organize and coordinate clinical trials of vaccines, promoting the development, production and incorporation of improved and new vaccines into national immunization programmes, and facilitating resources mobilization.

region and might include dengue, yellow fever, Japanese B encephalitis, and other locally important diseases.

The CVI integrates the efforts of numerous organizations and agencies throughout the world. The annual meetings of its Consultative Group are attended by professionals from many countries, representing governments, universities, research institutes, biotechnology and pharmaceutical firms, foundations, national and multilateral agencies, and special groups such as the Task Force for Child Survival.

Part of the work of the CVI is carried out through Product Development Groups (PDGs), which take on projects such as improving the heat stability of oral polio vaccine, designing a single-dose tetanus toxoid not given by injection and requiring fewer visits per pregnant female, and an improved measles vaccine that can be administered to younger infants. Technologies such as immunostimulating complexes, encapsulation, and timed-release products are featured.

Another major element in the CVI is a series of Task Forces, including the following:

- Priority Setting and Strategic Plans: to advise the CVI on priorities for action on new products, on the possible creation of new Task Forces or Product Development Groups and on regular updating of the CVI's Strategic Plan, and to monitor progress being made by other Task Forces and by Product Development Groups.
- Relations with Development Collaborators: to provide the CVI's Product Development Groups and the CVI Secretariat with guidelines for negotiating product development agreements with interested public- and private-sector parties.
- Needs Assessment and National Control Authorities of Developing Countries: to evaluate and, if necessary, strengthen national licensing procedures to assure the quality of vaccines used in immunization programs.

• Strengthening of National Epidemiological Capacities: to help countries strengthen their capability to collect, analyze, and use sound epidemiological data that are essential to the efficient incorporation of new and better vaccines into their national immunization programs; also to provide epidemiological services and identify important epidemiological issues for other CVI Task Forces and Product Development Groups.

Other Task Forces may be concerned with situation analyses of global vaccine capacities; assessment of the implications of introducing new and multivalent vaccines into ongoing immunization programs; technology transfer to developing countries; and logistic issues in vaccine handling, importation versus local production of vaccines, and similar issues.

A CVI Standing Committee, composed of representatives of the five sponsoring agencies, approves the strategic plans, programs, and budgets. A management Committee of about 25 members reviews policies, priorities, and program implementation. A much larger CVI Consultative Group consists of representatives of national development assistance agencies, international organizations, nongovernmental organizations, foundations, private and public sector institutions, commercial enterprises, and agencies and programs concerned with vaccine research, development, production, and delivery. The 1993 budget for all CVI activities was just under US$4 million.

Among the CVI activities has been a survey of global investment in research and development on children's vaccines. Information was gathered from national and international funding organizations, private and public sector vaccine manufacturers, foundations, and research councils. Total reported financial investment in 1992 was estimated at US$89 million, the great majority coming from public sector institutions. However, companies appear reluctant to divulge their specific research work, and private sector investment may be considerably greater.

Thus far the CVI has not devoted major attention to the integration of immunization programs with other aspects of primary health care, nor to alternative ways, such as water supply, sanitation, or education, to pursue similar goals in terms of health improvement.

*Integration within the WHO.* The six regional offices (in Alexandria, Brazzaville, Copenhagen, New Delhi, Manila, and Washington) are also involved in aspects of health technology transfer. For example, the Global Programme for Appropriate Health Care Technology, based in the European Regional Office in Copenhagen, has set up a Health Technology Assessment Network for the region. Within the Pan American Health Organization, the Advisory Committee on Health Research has a Subcommittee on Biotechnology that oversees a regional program in collaboration with UNDP, UNESCO, and UNIDO. The Inter-American Development Bank has also collaborated in helping to support regional conferences as well as a network of laboratories throughout the hemisphere. Examples of specific projects include synthesis of peptides, monoclonal antibodies, and DNA probes for diagnosis of malaria, Chagas' disease, and leprosy; studies of viruses of potatoes and other vegetables; genetic transfers for insect resistance in sugar cane and corn; and productivity of aquatic systems.

*A regional initiative, SIREVA (Sistema Regional de Vacunas) (Regional System for Vaccines).* This initiative is limited to the Region of the Americas, one of the six WHO regions. The Pan American Health Organization (PAHO), lead agency of the region, convened a conference in 1988 to discuss the topic, "Enhanced Disease Prevention in the Americas." This conference led to a strong endorsement of the need to increase investment in the field of vaccines. Accordingly, a study was undertaken by a group of vaccine experts supported by the government of Mexico, the Rockefeller Foundation, the InterAmerican Development Bank, the International Development Research Center (Canada), and PAHO. The group met in late 1989 in Mexico and enunciated a series of principles:

- Work should involve diseases relevant to Latin America and the Caribbean.
- Projects should involve different stages of work, from epidemiologic and laboratory research to clinical evaluation, pilot production, and field testing.
- Efforts should be made to draw from a wide range of disciplines and technologies and attract participation from a variety of host countries and affiliated laboratories in the region.
- The approaches selected should attract the attention of the international scientific community and the political and economic support of the countries of the region.

A Working Group from PAHO and the governments of Brazil and Mexico decided that these goals were workable and selected the following pathogens as initial priorities: *Streptococcus pneumoniae, Salmonella typhi, Neisseria meningitidis,* and dengue viruses. The existing PAHO coordinated laboratory network offers favorable conditions for these studies, with two regional laboratories, eight affiliated laboratories, and 15 epidemiological units. Other agencies are also cooperating. For example, the CDC will collaborate in determining the prevalence of different serotypes of *S. pneumoniae* in the region, the NIH is working with Mexico to transfer technology for a conjugated vaccine against typhoid fever, the U.S. FDA is collaborating on a meningococcal vaccine, the Walter Reed Army Institute of Research is working with Brazil on a dengue vaccine, and the Dutch-Nordic Consortium of Public Health Institutes has offered collaboration on a new pneumococcal vaccine. In addition, SIREVA intends to participate in other vaccine trials, such as that of new oral cholera formulations for use in the Region of the Americas.

As of early 1993 SIREVA remained in the planning stages as a regional approach to vaccine development that is complementary to the global programs outlined above. The appearance of similar consortia, based on the remaining WHO Regions or other geographic affinities, seems highly probable.

*United Nations Development Program (UNDP).* The UNDP funds technical assistance projects of all kinds in more than 150 developing countries, often in coordination with the other UN agencies listed. Within its "Global Programme" the UNDP, together with the FAO and the World Bank, established the Consultative Group on International Agricultural Research (CGIAR) in 1971. Members of the CGIAR now include the African, Asian, and InterAmerican Development Banks; the Arab Fund for Economic and Social Development; the European Community; the Rockefeller Foundation; and numerous national governments. The CGIAR network now extends to 13 laboratories in as many countries. One member, the International Laboratory for Research on Animal Diseases in Nairobi, studies trypanosomiasis and other protozoan diseases of livestock that have clear implications for human health. Many of the centers conduct research in plant molecular genetics and biotechnology.

The UNDP sponsors a Regional Biotechnology Program for Latin America and the Caribbean, and the Asian Network for Biological Sciences, both of which have strong interest in strengthening regional capabilities in biomedical aspects of biotechnology. Various country-specific programs are also supported by UNDP.

*United Nations Educational, Scientific, and Cultural Organization (UNESCO).* The work of UNESCO is relatively little known in the United States, which for political reasons discontinued its membership in 1985. This organization has a long interest in science and technology policy and in all aspects of technology transfer. Regional offices for science and technology in Amman, Jakarta, Montevideo, Nairobi, and New Delhi supplement the work of UNESCO headquarters in Paris. Two programs sponsored by UNESCO are of particular interest: the quadrennial conferences on the Global Impact of Microbiology and the broad network of Microbiological Resource Centers (MIRCENs). There are more than 20 participating MIRCENs in Africa, Asia, Europe, Australia, and the Americas, including several in the United States. The centers in Cairo, Guatemala City, Stockholm, Tucuman (Argentina), Kent (UK), Toulouse (France), Brisbane (Australia), and Trinidad and Tobago, have special concentrations in biotechnology (Caldwell 1988; da Silva and Sasson 1989). In addition, UNESCO supports the International Congress of Scientific Unions (ICSU), which has sponsored many biotechnology-related activities in developing countries through two committees. These are the Committee on Genetic Experimentation (COGENE) and the Scientific Committee for Biotechnology (COBIOTECH). The International Cell Research Organization (ICRO) is also sponsored by UNESCO.

*United Nations Industrial Development Organization (UNIDO) and the ICGEBs.* Based in Vienna, UNIDO is concerned with development of industries, including the pharmaceutical industry, in the poorer countries of the world. A special program has evolved planning and support for two International Centers for Genetic Engineering and Biotechnology (ICGEBs) located in Trieste, Italy, and New Delhi, India. The program has affiliated centers in many of the 45 member countries. The budget for 1992–1996 for the two Centers is US$72 million. The United States, Japan, and most western European countries have expressed neither interest in nor support for this program, which receives much of its financial backing from nonaligned and less-developed countries. Unofficially, the industrial countries appear to have doubts about the quality of the work in the two centers, and about free access to international markets in view of the high degree of intergovernmental collaboration present in the ICGEB program. All ICGEB member states have been asked to apply the "Voluntary Code of Conduct on the Release of Organisms in the Environment," which was formulated by UNIDO and approved by a Biotechnology Safety Working Group that included representatives from UNIDO, WHO, UNEP, and FAO.

The Trieste center specializes in regulation of DNA replication, molecular biology of papilloma virus, protein structure, and lignin degradation, while the New Delhi lab is interested in plant biology, immunology, molecular biology of hepatitis and malaria, design and synthesis of peptide antigens for disease resistance, and synthesis of oligonucleotide probes for DNA research. Research and training in New Delhi has been supported in part by the Rockefeller Foundation, and grants have been received from private industry in India. The two ICGEB centers became autonomous organizations on January 1, 1993. UNIDO publishes the *Genetic Engineering and Biotechnology Monitor,* which reviews progress in these fields throughout the world.

*United Nations Children's Fund (UNICEF).* This agency was created in 1946 as the United Nations International Children's Emergency Fund to help support the needs of children after World War II and was made permanent in 1953. In 1959 the UN General Assembly designated UNICEF as the agency responsible for carrying out the principles of the newly adopted Declaration of the Rights of the Child. Its support comes from voluntary contributions of governments (70%) and nongovernmental organizations and individuals (30%). UNICEF has become widely known as the chief architect of the "child survival and development revolution." Some agency priorities such as growth monitoring, oral rehydration, promotion of breastfeeding, female education, child spacing, and nutrition supplementation were discussed in chapter 2. Among its other programs UNICEF plays a leading role in integrating the provision and distribution of vaccines and immunization supplies for the EPI, and in helping to provide financial support and leadership to the other vaccine-related organizations listed here.

*Other UN agencies.* The Food and Agriculture Organization (FAO) supports many biotechnology-based projects that deal primarily with improvements in crops and livestock.

## Organizations outside the United Nations

The Organization for Economic Cooperation and Development (OECD), the European Community (EC), and numerous other regional groups as well as national organizations conduct training courses and symposia. Some of the many groups are listed below.

- Biosciences Information Network for Latin America and the Caribbean (BINAC), based in Caracas, supports information and bibliographic services for researchers in the region.
- European Federation of Biotechnology in Frankfurt conducts triennial congresses.
- European Molecular Biology Organization (EMBO), in Heidelberg, publishes the EMBO Journal and the proceedings of the many conferences and workshops that it sponsors.
- International Cell Research Organization (ICRO) in Paris was mentioned above as a project of UNESCO.

- International Organization for Biotechnology and Bioengineering (IOBB) in Guatemala City.
- Scientific Committee on Biotechnology (COBIOTECH), a product of UNESCO and of the International Council of Scientific Unions, provides information and advice to all countries about applications of biotechnology for the benefit of humanity. Its headquarters are in Santiago, Chile.

## National Efforts Toward Vaccine Development

*United States.* The U.S. government is heavily involved in supporting basic research in fields related to biotechnology, primarily through research awards from the National Institutes of Health and the National Science Foundation. A great deal of additional ''in-house'' research is done at laboratories of the Centers for Disease Control, the Food and Drug Administration, the Department of Defense, and other agencies.

Direct efforts in vaccines intended specifically for developing countries are undertaken through a collaborative program funded by the Agency for International Development (USAID) and administered by the National Vaccine Program Office (NVPO) of the U.S. Public Health Service with the advice of a Consultative Group on Vaccine Development (CGVD). This program provides funds for more applied aspects of vaccine work, such as the actual conduct of clinical and field trials.

The USAID also supports local scientists in collaborative research efforts with U.S. investigators in targeted vaccine R&D programs. An example is the Schistosomiasis Research Project in Egypt, which hopes to produce a vaccine against that disease.

*Other countries.* Many bilateral assistance and development agencies are involved in vaccine procurement and distribution, and some also support other vaccine-associated efforts. Programs and emphases change from time to time, and it is not possible to list specifics. The English-language names of some major national bilateral agencies are:

| | |
|---|---|
| Australia | Australian International Development Assistance Bureau (AIDAB) |
| Canada | Canadian International Development Agency (CIDA) |
| | International Development Research Centre (IDRC) |
| Denmark | Danish International Development Agency (DANIDA) |
| Finland | Finnish International Development Agency (FINNIDA) |
| France | No single agency; through Ministry of Foreign Affairs; Ministry of Economics, Finance, and Privatization; Ministry of Cooperation; Economic Cooperation Fund |
| Germany | Agency for Technical Cooperation (GTZ) |
| Italy | Department of Cooperation for Development (DCD) |
| Japan | Japan International Cooperation Agency (JICA); Japan International Development Organization (JAIDO, a mix of public and private) |
| Netherlands | Directorate General for Development Cooperation (DGDC) |
| Norway | Norwegian Agency for International Development (NORAD) |
| Sweden | Swedish International Development Authority (SIDA) |

Switzerland          Swiss Development Corporation (SDC)
United Kingdom       Overseas Development Administration (ODA)

## The Future of Immunization Programs

### *The Vaccines*

It seems certain that vaccines in widespread use, such as those in the EPI, will be further improved through technical changes very soon. For example, the pertussis component of DTP will be changed from the reactogenic whole-cell preparation to an acellular formulation with fewer adverse effects. Measles vaccine will be reformulated to be more immunogenic in younger infants, and oral polio will be made more temperature tolerant. Because all of these vaccines are officially sanctioned for universal application, their market is essentially assured.

It is also likely that new vaccines will be developed for diseases not currently immunizable, but even those that are safe and efficacious may not be marketable as individually administered products. One possible strategy for new vaccines for which there is little effective demand when used individually is to incorporate them into a "platform" such as DTP or MMR already in existence.

*When to start using a vaccine? When to stop?* The key questions for vaccine introductions are: How significant must a disease be to justify the use of a vaccine, or the implementation of a new vaccine-based program? Is there any objective test to define the product of prevalence and virulence for a disease that is widespread and relatively mild versus one that is rare but fatal? What is the "breakpoint" between the selective use of a vaccine for high-risk groups (e.g., by occupation, residence, genetic predilection, or personal behavior) and the decision to extend a vaccine to mass, universal, or compulsory application? When should a decision be local or regional, as with yellow fever vaccine; or global as with protection against hepatitis B?

The justification for routine use of a new vaccine in the United States was discussed by Rosenthal (1993) in relation to varicella (chickenpox). Almost everyone agrees that chickenpox is usually a relatively trivial childhood disease. Some who defend the new vaccine cite the cost of parental time for care, and possible loss of wages as sufficient reason for preventing chickenpox. Others refer to the possibility that following the illness, varicella virus can remain latent in nerve ganglia and reappear as shingles (herpes zoster). They would argue that the real intent of the chickenpox vaccine is to prevent more severe complications later in life. However, some scientists are concerned that the immunity induced in children may be only temporary. They are worried about the possible prolonged dormancy of the attenuated live varicella vaccine virus and the possibility of inducing the specific condition that the vaccine is intended to prevent:

> Could doctors in good conscience inject patients with a virus that would infect them forever? Would this mutant strain of the chickenpox virus be more likely to result

in shingles many years later? A miscalculation would be unfortunate, since shingles is far more painful and heals far less well than chickenpox in children (Rosenthal 1993).

The equation involving protective efficacy, risk of adverse effects, risk of the disease, cost, and convenience must be continually re-evaluated. Compliance falls off with perceived reactivity of the vaccine and with increased effort (repeat visits) or expense involved in obtaining immunizations. At some point, the numbers may argue against continued broad application of a particular vaccine, as has happened in the case of smallpox. Now poliomyelitis is considered a candidate for elimination. Assuming that this and similar optimistic predictions are borne out, policy decisions must some day be made by each national authority about the end-game strategy when an immunizable disease has become very rare but is not yet eliminated.

It is precisely in this phase of disease control that a vaccine backlash is likely to occur, because at some point the public will perceive, correctly, that the threat from the vaccine is greater than the threat from the disease. To the user, the safety of a vaccine is of greater concern than its efficacy, because deficiencies in safety are readily observable as adverse effects, whereas lack of protection is invisible. Continued willingness to be immunized depends on the perception of the danger posed by the particular disease and the risk that people will accept to prevent it.

As people become less tolerant of adverse effects, it will be more important than ever to preserve the integrity of the control program through a good-faith effort to compensate individuals who are, or believe that they are, injured by the adverse effects of the vaccine. Such a no-fault compensation program should be an integral part of any vaccine-based campaign, particularly when acceptance of the vaccine is compulsory. In the long run, it is likely that immunization programs cannot be continued indefinitely without some mechanism to compensate individuals and families for adverse effects.

It is important that the public understand the rationale for public health decisions, particularly when procedures appear to change. The idea in some countries of an end-game shift from live oral vaccine to inactivated polio vaccine to prevent sporadic vaccine-associated cases has been mentioned previously. The public must not view such a policy change as an admission of previous error but as a consciously planned and reasonable response to altered epidemiologic circumstances.

It is critical also that cost-conscious public officials not conclude that a formerly rampant disease made scarce by immunization has become inconsequential and spend their limited resources for other purposes.

*One disease, one vaccine, varying policies.* The threat from a disease, and the appropriate response, is not seen in the same light by all authorities. A good example is rubella, whose effects on children infected as fetuses during the first intrauterine trimester were discovered by the Australian ophthalmologist Norman M. Gregg in 1942.

Effective live viral vaccines, available since 1969, can immunize young females against infection during a future pregnancy and avert the risk of severe fetal damage. Nevertheless, as José and Olvera (1992) point out, most countries do not immunize against rubella. In some, this may be primarily for economic reasons, but many others have concluded that natural circulation of the virus, primarily in children, will assure that very few susceptible females reach childbearing age.

An alternative plan, called the "English strategy" by Anderson and May (1983), is used widely in Europe and in Japan, Australia, and Israel. This strategy calls for immunization of girls aged 10 to 14, with circulation of wild virus permitted. In areas of high incidence of pregnancy in young teenagers this policy would be dangerous and even counterproductive unless the age of immunization were lowered several years. A third system, called by Anderson and May (1983) the "American strategy," is to immunize all young children regardless of sex. For this purpose attenuated rubella virus is combined with measles and mumps viruses to make MMR vaccine, used in the United States and Canada with the intention of increasing herd immunity and halting the circulation of wild rubella virus.

José et al. (1992) and José and Olvera (1992), asked by the government of Mexico to advise on a national policy on rubella immunization, studied the seroepidemiology of rubella and recommended on the basis of their mathematical models that "the best strategy for preventing congenital rubella in Mexico is to continue with the no vaccination policy against rubella. Nonetheless, the continuous monitoring of this policy is necessary."

## Vaccines of the Future?

Up to the present time vaccines have been associated exclusively with infectious diseases, primarily in young children. However, continued expansions of vaccine technology, combined with increased knowledge of the human immune system, dictate a dramatic conceptual shift toward entirely new applications for immunization technology. The potential of immunoregulation of fertility has been mentioned previously. As countries pass through demographic and epidemiologic transitions (see Chapter 2) with a resultant change in the epidemiologic contour, vaccines may well be used primarily in adults and the elderly for control of chronic diseases. A glimmer of such an intention is seen in today's hepatitis B vaccine, utilized in large part to prevent liver cancer. Future vaccines may well target other malignancies. A vaccine against the Epstein-Barr virus (EBV) has been produced in the UK. This vaccine, whose main antigen is an EBV surface glycoprotein called gp340, is intended for use against nasopharyngeal carcinoma, which causes 50,000 deaths annually in southern China (Amato 1993). A vaccine against *Helicobacter pylori* might prevent gastric cancer. Vaccines might be applied against cardiovascular and other degenerative diseases such as multiple sclerosis. Is it conceivable to produce a vaccine effective against atherosclerotic plaques? Why not?

More is being learned all the time about the immune system, and immunopotentiating agents might be used in conjunction with vaccines to make them more potent and effective. As more is learned, immunogens can be designed to affect a particular target more specifically, or to induce secretory IgA differentially rather than IgG or other classes of immunoglobulins that are not especially effective against pathogens of the mucosal immune system. Cytokines such as interleukins 1 and 2 and gamma interferon have been shown to be effective adjuvants, enhancing protection induced by a variety of vaccines (Heath and Playfair 1992). Knowledge of the mode of action of the numerous regulatory hormones and cytokines can be applied to control the number and function of different types of normal and abnormal host cells and levels of soluble products, rather than against invading pathogens. At the same time, greater control is being gained over the responsiveness of the aging immune system and some progress is being achieved through use of dehydroepiandrosterone to rejuvenate immune function in old rats to make them more responsive to vaccines (Daynes 1992).

Could biological materials synthesized outside the body be used to prevent or ameliorate chronic metabolic diseases such as diabetes? Could an appropriately formulated and delivered vaccine protect against the carcinogenic effects of asbestos or other occupation and environmental agents able to penetrate human tissues? Will children one day be immunized against the adverse effects of air or water pollution?

In the future, orally or parenterally introduced materials will stretch the current concept of a "vaccine." Combined with specific nucleic acid sequences used in gene therapy, they may change the characteristics of individual humans (as well as animals and plants) in ways now imagined by only a few.

*The maximum likelihood estimate.* Speculations about future vaccines refer to pricey innovations not yet on the screen even in the wealthy countries, and light years away from the daily struggles for survival in the favelas of Rio de Janeiro, the bustees of Calcutta, or the alleyways of Nezahualcoyotl. Viewed in the spotlight of social equity, when so many people are unable to take advantage even of yesterday's technology, the rosy predictions of future wonders may begin to pale.

Are better vaccines through biotechnology a feasible path to freedom from disease, or, as some claim, merely an elitist tool to maintain dependency among the poorer countries? Are scientists naive when speculating outside their narrow field of expertise? There is no shortage of interpretations, but on balance there is justification for optimism, provided that

- the rate of world population increase continues to abate, leading to stabilization within the next few decades;
- the economic and environmental situations in the developing countries do not deteriorate further;
- anticipated improvements are made in the quality and variety of vaccines, in physical and economic access to them in the poorer countries, and in distribution systems;
- collaborations among scientists, industry, donor and regulatory agencies, and public health authorities at all levels are maintained and strengthened;
- comprehensive long-term support is given to the vaccine-related organizations described in this chapter, or to similar groups; and that the motivation and enthusiasm of the world community is maintained.

# APPENDIX A

# An Introduction to the Technologies

Those readers not familiar with the subject will find here a general description of some of the fundamental concepts and methodologies that underlie many modern biomedical innovations.

## Molecular Biology

Molecular biology, like all biological disciplines, is concerned with the study of living matter. Formerly discrete fields such as genetics, biochemistry, immunology and pharmacology converge in molecular biology. All share the goal of understanding the basic mechanisms, coded in genetic material, that control cellular differentiation and functions. As a scientific discipline, molecular biology originated in 1953, when Watson and Crick determined the double helical structure of deoxyribose nucleic acid, or DNA (Figure A–1).

### Genetic Engineering

Most efforts in genetic engineering apply the concepts and methods of molecular biology to modify living cells so that they will display particular features or produce desired substances. Recombinant DNA technology refers to the ability to transfer segments of genetic material from one organism to another. The controlled insertion of external genetic material results in partly synthesized cells that can manufacture complex biological materials. Two examples of such materials are human growth hormone and malarial sporozoite surface antigen. These cannot easily be obtained in large amounts by other methods, but can be synthesized rapidly, relatively cheaply, and in high purity by purpose-made genetically engineered cells. Alternatively, genetic engineers can try to fashion whole organisms with desirable characteristics not found in their native (or "wild") state. In this

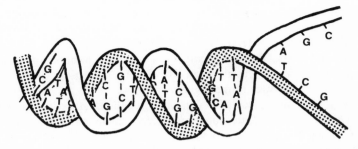

FIGURE A–1. A diagrammatic view of the double helical structure of DNA. The two invariant deoxyribose sugar-phosphate backbones are represented as thick strands, wound around each other and connected by hydrogen bonds from the four chemical bases or nucleosides that lie between them. Note that an A (adenine) always pairs with a T (thymine), in either order (A-T or T-A), and a C (cytosine) must pair with a G (guanine). This is called complementary base pairing. The strands are shown as linked on the left and separated on the right. Each separate single strand can define the order of base pairs on a newly formed strand that will form complementary to it, denoting the basis for DNA replication.

way, certain viruses and bacteria can be made to elicit immunity without inducing the illness that is being protected against.

Similar goals had of course been achieved many times before the invention of genetic engineering. Deliberate breeding over many generations has produced the familiar domesticated varieties of plants and animals whose procreation in the wild is normally self-selected. Useful microorganisms such as yeasts and bacteria have been cultured, selected, and improved for thousands of years.

The difference between selective breeding and genetic engineering is in the capability of the latter to identify, select, isolate, transfer, and express specific genes of known structure and properties. In this way not only can "natural" evolution and selection be accelerated, but functional and genetic capabilities can be transferred from one species to another in a way that would never happen in nature.

To accomplish these tasks, genetic engineers depend on a thorough knowledge of deoxyribonucleic acid (DNA), which controls all aspects of cellular behavior and predetermines the characteristics of each individual organism. Despite the complexity of the information encoded within it, DNA itself has a fairly simple molecular structure. It consists of a backbone of sugar (deoxyribose) to which one of four different bases (adenine, cytosine, guanine, and thymine) can attach at regular intervals. The order and arrangement of these bases acts like a four-letter alphabet to "spell out" the code that controls cell functions. Each sequential set of three bases represents a different one of the twenty common amino acids that make up all of the peptides and proteins, including enzymes, that make up all cells and regulate biochemical reactions. In essence, a gene is a sequence of DNA that encodes for a single protein. Many other DNA sequences precede and follow genes and control those proteins that regulate the way in which the information within the DNA is used to synthesize "its" protein (gene expression). Remarkably, the basic scheme to accomplish this purpose is the same in all known organisms, including viruses, bacteria, yeasts, other plants, and animals from amoebae to mammals. However, the complexity of the genetic information (and of the resultant products), as well as the methods of gene regulation, become increasingly greater from the lower to the higher organisms.

Much early work in molecular biology and genetic engineering has made use of *Escherichia coli (E. coli)*, the common bacterial inhabitant of the human colon. At present, although much work continues to be done with *E. coli,* a great variety of other bacteria, yeasts, and plant and animal (including human) cells is used for the study and transfer of genes.

*Mechanics of gene splicing.* How does a genetic engineer go about inducing a cell to make a particular protein? There are various protocols by which this process can be accomplished, but in general the following steps are included:

1. Using his or her knowledge of gene regulation and expression, the genetic engineer obtains a sequence of DNA, either from another organism or by chemical synthesis, that contains the coded information to permit the cell to manufacture a useful protein. Examples of such proteins are human insulin, growth hormone, hepatitis B surface antigen, and the blood-clotting protein factor VIII, absent from the blood of hemophiliacs. With the aid of restriction enzymes (biologically active enzymes that cleave nucleic acids only at a particular sequence of bases) the source DNA molecule is cut into pieces.

2. Appropriate pieces of the nucleic acid are inserted into a vector, which is usually a bacteriophage (also simply called a phage) or a plasmid. A bacteriophage is a virus that infects bacteria; in the case of *E. coli,* lambda phage is the most widely studied and used of the phages. A plasmid is a circular piece of DNA, found commonly in bacteria. The insertion is done by using restriction enzymes to open the vector's own DNA, introducing the piece of DNA that is to be transferred, and then using enzymes called ligases to join segments of previously cleaved DNA. Both plasmids and phages contain regions of DNA known as promoters, which initiate and maintain high levels of transcription (DNA-directed protein synthesis). By splicing the sequence of DNA encoding the desired protein into a region of the plasmid or phage DNA immediately following the promoter, the genetic engineer ensures that the bacterium will produce a large number of copies of the encoded protein.

3. When the appropriate plasmid or phage has been constructed, the genetic engineer introduces the vector into colonies of bacteria or other host cells.

4. The host cell colonies are then screened to identify those that have successfully taken up the vector and are expressing its genetic instructions.

5. The genetic engineer must then isolate and grow large numbers of those cells that express the protein.

6. The final step is concentrating, purifying, and verifying the sought-after protein from the thousands of other components in the cell and culture medium.

Despite the intricacy of the procedures leading to a genetically engineered product, this technique has been used to produce hundreds of biologically active proteins. As a testimony to its success, the four proteins cited above (insulin, growth hormone, hepatitis B surface antigen, and factor VIII) have been produced through genetic engineering and are all available commercially. The hepatitis antigen serves as the basis of a successful vaccine.

Sometimes the genetic engineer is not interested in obtaining the protein product encoded by the DNA sequence, but in making numerous copies of the DNA itself. Two reasons to accumulate specific DNA are to analyze the composition of a gene being studied, or to make hybridization probes, as described in a subsequent section.

If it is in fact not a particular protein that the genetic engineer seeks to produce but a novel organism, a different set of methods is employed, although all still depend on the fundamental processes of DNA transcription and regulation. With regard to medicine, the construction of novel organisms finds its greatest application in the development of new vaccines. Viral vaccines, discussed in Appendix B, show the greatest promise as genetically engineered whole organisms for immunoprophylaxis. By deliberately introducing modifications into the viral DNA or RNA, researchers hope to produce attenuated viruses that retain all the immunogenicity (the ability to generate an immune response, and ultimately immunity, in a host) of the disease-causing virus while eliminating the harmful effects that normally accompany viral infection. In addition, work is under way to insert DNA that encodes the immunogenic surface proteins from disease-causing viruses into the DNA of harmless viruses, thus eliciting an immune response to the disease-causing organism without exposing the host to the actual pathogenic agent. Although theoretically the principles outlined above should permit the construction of novel bacterial strains that induce immunity without leading to disease, the far greater complexity and number of bacterial genes has hampered progress in this arena.

## Immunologic Methods: Humoral Immunity

### Antigens

Since the beginning of scientific immunology it has been known that vertebrates, including mammals, produce specific chemical substances in response to introduction into the body of certain foreign materials. These materials may be naturally occurring living microorganisms that penetrate by themselves or via insect vectors, allergens such as pollens, or nonliving antigenic molecules. Often these are proteins, but other materials such as polysaccharides are antigenic under certain circumstances.

### Immunoglobulins (Antibodies)

The soluble substances that appear in the blood after antigenic stimulation became known as antibodies or immunoglobulins. They were found to be complex proteins containing a subregion that binds specifically to a distinct site ("epitope") on the evoking antigen.

Over the years, many ideas were proposed to explain antibody production and activity. Early researchers such as Paul Ehrlich conceived the idea that the antibody and its inducing antigen fit together physically like a lock and a key, a concept that has in general been confirmed. The biological basis for the generation of diverse antibodies was a mystery until 1959, when the Australian immunologist Macfarlane Burnet envisioned the theory of *clonal selection,* for which he was later to receive a Nobel Prize. Burnet believed that a great variety of different, potentially antibody-producing, cells exist in every normal human or higher animal. The existence of these cells, belonging to the B-lymphocyte class of white blood cells, has now been established without question. Each such cell, when activated to become a plasma cell, has the capability to secrete a single specific configuration of antibody. The cell is called into action when its particular evoking antigen enters the body and is recognized. The resultant stimulation causes that cell to differentiate and proliferate, leading to a *clone* of identical daughter cells. All cells in this clone secrete the same distinctive antibody capable of forming a highly specific chemical bond with the antigen that stimulated its production. The secreted antibody enters the bloodstream, where it mixes with the diverse immunoglobulin products of all other activated clones of antibody-producing cells.

Humans can make five different classes of immunoglobulins. Each immunoglobulin (Ig) molecule consists of four arrays of amino acids that form two heavy and two light polypeptides chains. The chains contain variable regions that encode the specificity and constant regions that are the same for all immunoglobulins of that particular class. The classes differ from one another principally in the constant region of the heavy chains. The five classes, in decreasing order of abundance in serum, are: IgG (formerly called gamma globulin), IgA, IgM, IgD, and IgE. The molecular weight of IgG is about 146 kD (kiloDaltons), which permits antibodies of this class to cross the placenta and to protect newborns against many kinds of infection. By contrast, the molecular weight of IgM is about 900 kD, which is too large to cross the placental blood vessels.

Detectable immunoglobulins of the various classes appear in the blood at distinct times after antigenic stimulation and persist for differing periods of time. IgM is made early in infections, but IgG tends to be longer-lived and may persist for extended periods. IgA is found primarily in secretions such as saliva, tears, and intestinal, genitourinary, and bronchial mucus. It helps to prevent the attachment of cholera vibrios, *Shigella,* and *E. coli* bacteria to the intestinal epithelium. IgE is associated with allergic reactions. Although much is already known, the full range of functions of these essential molecules remains to be worked out.

One of the most important functions of immunoglobulins is to serve as antitoxins; that is, to neutralize the toxins produced by certain bacteria, such as streptococci or the agents of diphtheria or pertussis. The ability to form specific antibodies to bacterial toxins is the basis of immunization against diphtheria and tetanus, whose toxins are modified by treatment with formaldehyde or precipitation with alum to form immunogenic but nonpathogenic toxoids. Acellular toxoid-based pertussis vaccines are used in certain countries and recently have been licensed in the United States. Immunoglobulins can also destroy certain viruses and bacteria as well as protozoa such as trypanosomes. They can block the attachment of viruses and bacteria to various cells and prevent the invasion of red blood cells by malarial parasites. Antibodies also have complex effects in stimulating the attachment of eosinophils and macrophages to certain worm parasites such as schistosomes and nematodes, and the subsequent killing of the parasites.

*The antibody response to immunization.* In assessing the humoral response to immunization, the specific Ig class, the degree of antibody affinity to its evoking antigen, and its epitope specificity are more significant than the overall immunoglobulin titer. Nonspecific Ig is made as a response to various infections. Some specific antibodies have little or no protective function because they are reactive to parts of organisms not involved with toxins, attachment, invasion, or other aspects of virulence. Therefore the level of immunoglobulin, or even of specific antipathogen Ig that can be detected in tests such as ELISA may be useful for diagnosis but is not necessarily correlated with protection against future attacks.

*The polyclonal pool.* The types and amounts of antibodies in the blood of an animal or human at any moment consists of a vast variety of immunoglobulins and represents, more or less, the cumulative responses to antigenic stimulation over the lifetime of that individual. It is now know that the attributes of the B lymphocytes, including their potential to respond to particular antigens, are characteristics inherited with the major histocompatibility complex, and vary from person to person. Therefore not all individuals, or populations of individuals, who are exposed to a particular antigen will react to it in the same way. Such differences are crucially important for deployment of vaccines and are discussed further in Appendixes B and C. The immune system is also subject to modulation through external and internal life experiences of individuals, for example by aging, nutritional condition, or certain infections of which acquired immunodeficiency syndrome is the most spectacular.

*Monoclonal antibodies.* For decades immunologists had obtained antibodies for their research by injecting foreign antigens, either crude or purified, into rabbits or other animals. After some time the rabbit was bled, its blood was allowed to clot, and the clear serum, containing the pool of the animals' total immunoglobulins, was collected. Individual rabbits, like animals of all other species including humans, vary in their innate ability to recognize and respond to particular antigens. Consequently, the antibodies produced varied from rabbit to rabbit or even within the same animal over time. For laboratory, diagnostic, or experimental purposes, the principal disadvantage of polyclonal serum is the inherent inconstancy of its immunoglobulin composition. The inability to obtain a uniform, reproducible, but above all, specific product hindered biomedical research and practice.

When the existence and immune functions of B-lymphocyte cells were recognized, many immunologists attempted to isolate and grow them in artificial cultures, but the cells had only a limited life span and stubbornly resisted propagation *in vitro*. However, some other cells, such as certain malignant tumors of the immune system (myelomas), are capable of continuous proliferation in cultures. Unfortunately, those cells produce little or no antibody. In the mid-1970s, Köhler and Milstein, at Cambridge University, devised a method to fuse together the immortal but nonsecreting myeloma cells and the secreting but nonculturable lymphocytes to produce hybrids with the desirable properties of both parents. This is the "hybridoma" (hybrid myeloma) so commonly used today.

Individual fused cells can be isolated in nutrient medium, in which they will grow and multiply to produce a clone of identical cells that can be propagated in theoretically infinite numbers. All cells will secrete a structurally uniform immunoglobulin, the *monoclonal antibody*.

A monoclonal antibody is a well-defined chemical reagent that can be produced in volume, in contrast to a polyclonal antiserum from an intact animal, which is a variable mixture of reagents that is never duplicated precisely from one batch to another. A substantial industry has sprung up to make monoclonal antibodies, primarily as laboratory reagents. Because of their specificity, monoclonal antibodies can be used to identify particular molecules and immunoreactive sites, providing the basis for many different diagnostic tests. In theory, any kind of antigenic material can be caused to bind with its own complementary monoclonal antibody.

Among the practical applications of monoclonal antibodies are: diagnosis of an infection and identification of the particular type or strain of pathogen; use as "probes" for defining the antigenic composition of organisms or their products; identification of the specific subunits within an antigenic complex that are most likely to stimulate immunity (e.g., for a possible vaccine); standardization of diagnostic reagents and vaccines; and possibly as a vehicle to deliver drugs or other means of immune-based treatment to specific target sites.

*Single-domain antibodies (dAbs).* The production of monoclonal antibodies to obtain pure immunoglobulins is a tedious, complex, and costly procedure. Recent advances in biotechnology have wedded immunology to molecular biology for the production of even more specifically combining molecules. In the case of single-domain antibodies, only the variable region of the immunoglobulin molecule is duplicated, containing the binding sites that confer specificity. This is done by isolating the DNA sequence that codes for production of that particular variable region, inserting that gene into *E. coli* bacteria, and harvesting the expressed polypeptide from the culture medium (see the preceding section on molecular biology techniques).

Single-domain antibodies have the advantages of small size and exquisite specificity and can be made by bacteria in fermentation vats, avoiding the complexities and hazards inherent in mammalian tissue culture. However, technical problems may cause difficulties in production and application. The invariant (constant) region of the immunoglobulin molecule is the same for all antibodies of one class. Because the constant region is not produced, the single-domain antibody represents only a portion of the native immunoglobulin. Therefore it cannot be used in diagnostic tests such as indirect immunofluorescence or enzyme-linked procedures, described below, that depend on the presence of the entire immunoglobulin molecule.

## Immunologic Methods: Cellular Immunity

If it is recognized that immunoglobulins and other humoral factors (such as complement) are secreted by cells, then all immune reactions are cell-mediated. Moreover, acquired resistance can be demonstrated in the absence of detectable antibody, as shown by local inflammatory reactions, and activation and proliferation of specific cell types. The intact immune system is made up of many cell types, all of which are derived eventually from pluripotent stem cells in the bone marrow. The general categories of immune effector cells are:

1. *Polymorphonuclear leukocytes,* which constitute 60% to 70% of circulating leukocytes (white blood cells), occur in three types. Under normal circumstances about 90% of these cells are *neutrophils,* which are phagocytic and remove invading organisms such as bacteria. In certain parasitic infections, for example, *eosinophils* may constitute a very high proportion of white blood cells. *Basophils* are relatively fewer in number and poorly understood.

2. *Monocytic cells* pass from the bone marrow to the blood and then to tissues, where they convert to macrophages. As the name implies, macrophages are specialized to ingest and remove viruses, bacteria, protozoa, and certain nonliving materials. These cells play a major part in defending the body from infection. They secrete various hormone-like monokines, including cachexin (tumor necrosis factor) and interleukins, that direct the activities of other parts of the immune system. Monocytic cells in turn are affected by other humoral and cellular components.

Stimulation of macrophages may be accompanied by changes in size, shape, and contents and a greater propensity to specific or nonspecific killing of certain microbes. The very complex system of chemotactic and stimulatory cytokines, interleukins, lymphokines, and other hormone-like factors which are produced by lymphocytes and other cells cause recruitment of macrophages or other effector cells that may interact with IgE to kill invading pathogens.

3. *Lymphocytic cells* may take either of two developmental paths, differentiating in the thymus to become T-lymphocytes, or in the bone marrow as B-lymphocytes. Primordial T-lymphocytes may differentiate further as effector cells, active in cellular immunity, or as regulator cells, which in turn may become T-helper or T-suppressor cells. When activated, the B cells mature to plasma cells, which actually secrete immunoglobulins.

Purified T-lymphocytes can inactivate certain viruses, produce interferons with antiviral action, and have other cell-mediated protective functions.

Lymphocytes secrete the hormone-like lymphokines that play various roles including activation of B cells. Many different kinds of lymphokines and related materials, such as interferons, have been described, and more are undoubtedly waiting to be discovered. These informational molecules are critically important in orchestrating and regulating host immunity, particularly against parasites, and their functions must be understood if a new vaccine is proposed.

## Diagnostic Methods

### Immunologic Methods

*Serology.* Serologic diagnosis of infectious diseases requires a sample of host immunoglobulins (antibodies), usually in blood serum, and a source of antigens specific to the pathogen(s) being tested for. Classical or conventional methods of immunodiagnosis, even in competent hands, often require considerable experience for interpretation. Technical factors such as composition, concentration, potency, and uniformity of reagents and buffers, training of laboratory personnel, and calibration of instruments all affect the reportable results of serologic tests. This is particularly true in smaller laboratories that perform fewer tests.

In evaluating vaccine immunogenicity or efficacy it is critically important to obtain reliable antibody titers from the same individual before and after immunization. Other diagnostic protocols call for serology at different stages of infection, such as incubation, illness, or convalescence. Such studies, as well as seroepidemiologic surveys, demand a high degree of precision, standardization, and reliability. Consistent accuracy requires quality-control measures such as periodic testing of blind-coded test samples of known composition to confirm the accuracy of results.

Recent advances in *in vitro* culture techniques provide a source of purified fractions of various pathogens for use in diagnostic testing. In addition, recombinant DNA technology and well-controlled chemical synthesis can produce highly specific proteins and peptides for this purpose.

The logical antithesis of serologic tests (which measure host-produced antibodies) is the detection of circulating or tissue-bound antigens that come from an infecting organism. Whereas serology may demonstrate an ongoing or past infection, the presence of pathogen-derived antigens usually indicates current, or at least recent, infection. The typical immunologic test for a specific antigen includes a stage during which monoclonal antibodies, or polyclonal sera containing the specific antibodies, are permitted to react with the patient specimen, followed by washing to remove unreacted immunoglobulin, and finally a procedure that provides a signal that the specific binding has taken place. This signal may be fluorescent, enzymatic, or even magnetic. A common configuration for such a procedure is an enzyme-linked immunosorbent (ELISA) assay.

Immunologic tests for antigen detection may compete with molecular methods including nucleic acid amplification methods and hybridization probes, which can identify pathogens by their chromosomal or extranuclear nucleic acid base sequences. These molecular methods are described later in this Appendix.

The basic kinds of immunologic tests are:

1. *Agglutination tests* based on the clumping together of small particles by the combination of antigen and antibody. In the *direct agglutination assay,* whole test organisms (usually grown in the laboratory) are mixed with the unknown serum. Clumping of the organisms indicates the presence of a specific antibody in the serum. Examples are the Widal test for *Salmonella* and direct agglutination assays for malaria parasites and for *Trypanosoma cruzi,* the agent of Chagas' disease. In *indirect* or *passive agglutination,* serum antibodies cause clumping of antigen-coated cells or inert particles such as bentonite (a mineral) or tiny latex beads. Red blood cells, commonly from sheep, are used in the *indirect hemagglutination assay. Indirect hemagglutination inhibition* is a competitive assay between soluble antigens and antigens bound to particles, in which the agglutination of particle-bound antigens is inhibited by the presence of antigens in the sample.

Agglutination tests are relatively simple and are sufficiently sensitive for many routine purposes. They do not require costly instruments and can be performed in a simple field laboratory. On the other hand, they are variable and difficult to standardize and are relatively inefficient for the IgG class of antibodies. Nonspecific interference from rheumatoid factor and certain antibodies can lead to false positive reactions.

2. *Precipitation* tests depend on the formation of insoluble complexes when antigens and antibodies react, producing visible lines or bands is semisolid substrates such as agar or agarose gels. One example of a precipitation test is the Ouchterlony double diffusion test, in which antigen and antibody are placed in separate wells in a gel and diffuse outward, encountering each other and forming a line of precipitation. Another example is immunoelectrophoresis (IEP) and counterimmunoelectrophoresis (CIE), in which the reagents are caused to migrate toward each other by an electric current. IEP and CIE procedures are relatively fast and simple and require only a basic laboratory with an appropriate power supply. The CIE test has been used in diagnosis of a variety of viral, bacterial, parasitic, and fungal diseases.

3. *Complement fixation* (CF) is the oldest, and formerly the most widely employed, of all tests to determine the presence of antibodies in serum. The CF test is based on the activation of blood complement components by antibody-antigen complexes. The CF test has been used for detection of numerous infectious and parasitic diseases. It has good specificity, but it is technically and conceptually complex, and quality control is difficult. The reagents are variable, and complement itself is relatively costly, unstable, and has a limited shelf life. Equipment such as centrifuges and refrigerators may present a problem in the field, and practical problems in reagent storage conditions as well as technician training and monitoring may limit the application of CF in developing countries to well-equipped, specialized laboratories. The classical CF test is being replaced by more modern methods.

4. *Immunofluorescence* (IF) has been used in diagnostic tests since the 1940s. Antibodies or antigens are conjugated (permanently bound) to chemical fluorochromes, which fluoresce brightly under ultraviolet (UV) light. A microscope with a UV light source and special optics is required for immunofluorescent observation of tissue sections on microscope slides. Immunofluorescence assays may be direct or indirect. (1) In the direct (DIF) test, a tissue section is incubated with specific antibody that has been conjugated to a fluorochrome. After excess unbound antibody is washed away, the location of the desired antigen in the tissue is indicated by fluorescence of the bound fluorochrome-labeled antibody. (2) The indirect (IIF) test uses an unlabeled first antibody, for example, a mouse-derived monoclonal that is specific to the antigen being sought. After this specific antibody is allowed to react with the tissue (and the excess is washed away), the preparation is incubated with a second antibody that has been conjugated to a fluorochrome. The second antibody, which was made in a rabbit, goat, or other species, is directed against mouse immunoglobulin. The resultant ''sandwich,'' visible under the UV microscope, signals the sites where the monoclonal antibody attached to its specific antigen in the tissue and the labeled antibody then attached to the monoclonal.

An IIF test can also be used to detect the presence of a specific antibody in serum. Here, antigen is immobilized on a solid phase such as a plastic surface. The patient serum is added and incubated for a certain period. The unbound serum is washed away and fluorochrome-labeled antibody against human immunoglobulin is added. After another washing, ultraviolet illumination reveals the sandwich of [antigen + patient antibody + conjugated anti-human Ig], demonstrating the presence of the antibody in the patient's serum.

Immunofluorescent methods are used to diagnose many infectious viral, bacterial, and parasitic diseases. The growing availability of specific monoclonal antibodies is helping to make viral immunoassays more sensitive and specific.

One difficulty with IF tests is that the fluorescence fades rapidly and a permanent record can be made only by photography.

5. *Enzyme immunoassays* (EIA) are similar to immunofluorescence except that they employ an enzyme rather than a fluorochrome as an indicator to signal the occurrence of an antigen-antibody reaction. EIAs have become an essential tool for immunodiagnosis of many infectious and other diseases.

An EIA uses an enzyme conjugated either to an antigen or to an antibody in a way that preserves the catalytic function of the enzyme and the immunoreactivity of both reagents. The pattern of incubations and washings is similar to that in immunofluorescence, except that in place of UV light there is a final incubation with a chemical substrate that produces a colored product in the presence of the enzyme. Generally peroxidases and phosphatases are used, and the color reaction is stable and permanent, in contrast to the liability of immunofluorescence reactions.

The most important EIA is the ELISA (Enzyme-linked immunosorbent assay), in which one immune-reactive element (usually the antigen) is attached to a solid-phase surface such as plastic beads, disks, or tubes. The concave wells of molded plastic microplates are often used, in which 48 or 96 tests may be performed simultaneously. A small amount of diluted patient test serum is added to each well. After interaction under appropriate conditions, and washing as described above, an enzyme-labeled anti-human immunoglobin is added. After a second incubation and washing, the appropriate substrate is added and enzyme activity in the well is measured by determining the degree of resultant color. ELISA tests, so important in diagnosis of AIDS, are commonly conducted and read by automated equipment.

ELISA tests have been used for antibody measurement in virtually all human and animal infections. A single serum sample may be divided and tested for the presence of any number of antibodies, using the identical enzyme-linked anti-immunoglobulin reagent, provided only that the various antigens are bound to separate wells of the microplate(s).

There are many variations of ELISA and other EIA tests, all of which require (1) a chemically correct and properly prepared solid phase, (2) suitably diluted (or concentrated) samples of both antigen and antibody, (3) chemically active enzyme and substrate, (4) fresh diluents and buffers of the proper chemical composition, (5) a correctly timed sequence of incubations and rinses, and (6) a sensitive detection mechanism. Enzyme-linked immunoassays are extremely sensitive and highly adaptable, especially when configured with monoclonal antibodies for antigen detection. However, they can be complex and time consuming and quality control must be carefully regulated.

6. *Immunocytochemical tests* are used to demonstrate the presence of substances, usually antigens, in tissue sections. Both immunofluorescent and enzyme-linked methods can be applied to frozen or paraffin-embedded tissue sections. When the enzymatic procedures described above are used, a permanent color in the tissue section shows the precise location of the antigen being sought. When fluorescein-conjugated antibodies are used, a bright apple green to yellow fluorescence, visible through a fluorescent microscope, shows the same areas. Monoclonal antibodies increase the specificity of immunocytochemical procedures.

7. *Immunoblots or "Western Blots"* (WB) are much used in confirmatory tests for HIV antibodies following ELISA. Immunoblots are made by first preparing a polyacrylamide slab gel in which the peptides of the sample antigen are separated by an

imposed electric current into a series of parallel bands according to molecular weight. A piece of nitrocellulose paper is then held firmly against the broad surface of the gel and, under the influence of another electric current, the separated protein and polypeptide bands are caused to move out of the gel and onto the paper, where they become firmly bound. Subsequent incubation of the paper with antibodies against one or more of the bands will result in an immunological reaction at the suitable places on the paper. Incubation of the paper in an enzyme-conjugated second antibody and a solution of the appropriate enzyme substrate, as described above, will result in deposition of a colored product at the precise site(s) of the antigenic band.

The preliminary fractionation of the antigen sample and subsequent incubation with patient serum permits identification of different antibodies in the serum that are reactive with specific bands. The WB can therefore differentiate subsets of antibodies that would all give an apparently identical reaction by ELISA or immunocytochemistry. However, immunoblotting is complex and expensive and requires specialized apparatus and chemicals as well as trained technicians.

8. *Radioimmunoassay (RIA)* is a powerful but demanding procedure used to determine levels of hormones, drugs, and many antigens in a variety of samples, or to determine the level of antibody in serum. For that application, the known antigen is immobilized on a solid phase such as tubes, wells, or particles; the sample serum is added; and a specific antibody is permitted to attach. The bound antibody is detected and quantified with a radiolabeled antiimmunoglobulin that attaches to the bound antibody. Radioimmunoassay is a highly sensitive procedure that can detect as little as one picogram (one trillionth of a gram) of antigen. Such precision may be greater than is needed for a diagnostic determination. The method is expensive and usually requires radioactive isotopes, principally $^{125}I$ and $^{131}I$ that have relatively short half-lives and are gamma emitters. Beta emitters such as $^3H$ or $^{14}C$, which have very long half-lives, are sometimes employed. Quantitative RIA tests require a gamma counter for gamma emitters, or a scintillation counter for beta emitters. The shipment, use, and disposal of radiochemicals raises serious legal and environmental issues. These and other technical problems preclude the use of RIA as a routine test in the field in developing countries.

## Collection of Blood for Testing

Venipuncture (phlebotomy) is the standard method for obtaining quantities of blood sufficient for serologic testing. Micromethods can use blood collected in capillary tubes from finger sticks. In infants and young children a needle stick in the heel or the earlobe will produce a few drops of blood with little trauma. In lieu of glass tubes or vials, many investigators have employed filter paper to absorb a few drops of blood. The dried filter papers can be stored for several months in a cool and dry environment away from insects and fungi. In the laboratory, the blood spot can be eluted from the filter paper, and some immunodiagnostic tests can be performed directly on the paper.

## Assays for Cell-Mediated Immunity

In addition to the tests just described that estimate the degree of antibody production (i.e., humoral immune reaction) in response to an antigenic stimulus, several methods are in common use to evaluate cellular responses. The simplest is a differential count of the various classes of white blood cells including polymorphonuclear leukocytes (neutrophils, eosinophils, and basophils), mononuclear cells of various types such as monocytes and macrophages, and several subsets of lymphocytes. Both absolute number and percentage distribution are significant.

Assays on polymorphonuclear cells and macrophages include tests of phagocytic and bactericidal ability, chemotaxis, and production and release of several enzymes. For lymphocytes, uptake of radioactive thymidine may be measured in the presence of substances (mitogens) known to stimulate cell proliferation. The labeled thymidine is incorporated into DNA, providing a direct measure of cell division. The amount of division is related to the degree of cell activation and other characteristics of interest and provides information about previous experience with certain antigens. For all categories of leukocytes, surface receptors, membrane-bound antigens, and other biochemical markers may be identified by use of commercially available specific monoclonal antibodies.

Some tests of white blood cells are done by microscopy, and some require cultivation for one or more days in specific media in multiple well plates, tubes, or dishes. The most elaborate method of study involves the use of the fluorescent activated cell sorter (FACS), sometimes called a flow cytometer. This complex machine automatically measures the intensity of laser-stimulated fluorescence emitted by cells that are sent in a continuous stream past a sensor. The FACS sorter is able to distinguish different cell types and identify those having particular membrane or internal markers. The machine also tallies the numbers of each type of cell for which it is programmed. Needless to say, the more sophisticated assays of leukocyte type and function require an abundantly equipped laboratory and highly trained staff. These procedures are beyond the limits of achievement in all but the most highly evolved reference laboratories in developing countries.

## Molecular Methods of Diagnosis

### Nucleic Acid Hybridization Probes

The major contribution of molecular biology to the diagnosis of disease agents is the nucleic acid hybridization probe and its enabling technologies. This powerful tool is useful also for diagnosis of certain malignancies and genetic diseases and for some forensic applications. Probes based on DNA can, at least in principle, function on material years or centuries old and unsuited to delicate immunochemical manipulation, and on many organisms including:

> Mycoplasmas; bacterial endotoxin genes; latent viruses such as herpes, CMV, HIV; human papilloma viruses; chronic hepatitis B in liver tissue; direct identification of antibiotic resistance genes in bacteria; detection of virulence determinants such as the virulence plasmid in *Yersinia pestis; Helicobacter* in stool; *Plasmodium falciparum* (Kingsbury 1987).

Nucleic acid hybridization is analogous to immunologic methods in the sense that a prelabeled component is used to attach chemically to a highly specific target configuration in a kind of "lock-and-key" arrangement (Figure A-2). Rather than an antigen-antibody reaction, DNA probes rely on the mutual chemical binding of complementary DNA (or sometimes RNA) sequences found in all cells and can even detect certain viral nucleic acid sequences that have become integrated into host cell genomes.

*Theoretical basis.* As mentioned earlier, the DNA molecule contains four chemical bases or nucleosides: adenine (A), cytosine (C), guanine (G), and thymine (T). Like the rungs of a ladder, pairs of nucleosides connect the two deoxyribose-phosphate strands, but an A can fit only with a T, and similarly a C must pair with a G to fit exactly between the two strands of the molecular backbone. Therefore, knowing the sequence of bases on one strand, say A-T-A-C-G-T, one can deduce the complementary sequence T-A-T-G-C-A that must be present on the other strand if they are to fit together correctly.

Hybridization procedures depend on this complementarity, using specifically created probes that contain unique genetic sequences present in the organism to be tested for. In one method, restriction endonuclease enzymes are used to cleave DNA at predetermined sites, and the DNA fragments so produced are separated by electrophoresis. The selected fragment is then copied many times by special amplification techniques described below, and labeled by incorporating a radioactive isotope or attaching some other marker. Alternatively, if the base sequence is known, a DNA synthesizer can be used to make the complementary probe, which can then be similarly labeled.

*Procedure.* In a hybridization procedure, the two strands of the sample DNA are separated (denatured) and attached to a solid phase such as a nitrocellulose membrane or a glass slide. The labeled DNA probe is denatured by heating and is added to the sample. The preparation is incubated and then washed to remove unreacted probe, leaving the probe strands containing complementary DNA sequences firmly bound to the separated strands of sample (target) DNA. This binding is detected via the label that was attached to the probe. A radioactive isotope incorporated into the label can be localized by autoradiography; that is, by placing photographic film in contact with the preparation (in the dark, of course) for a sufficient length of time, and then developing the film and searching for darkened areas. Other labels depend on a color reaction similar to those described previously for ELISA tests.

In a *Southern Blot*, the DNA in the sample is first cleaved by endonucleases into fragments. These pieces are separated by electrophoresis, denatured into separate single strands, and transferred (blotted) by electrophoresis to a nitrocellulose membrane. Hybridization is performed as described above, and individual DNA bands to which the probe has attached are revealed by the standard detection methods such as autoradiography.

*In situ* hybridization permits the detection of DNA sequences directly in tissue biopsies or other histological sections, showing precise localization of the target DNA sequence. For example, liver cells containing hepatitis virus or red blood cells containing malaria parasites can be identified precisely.

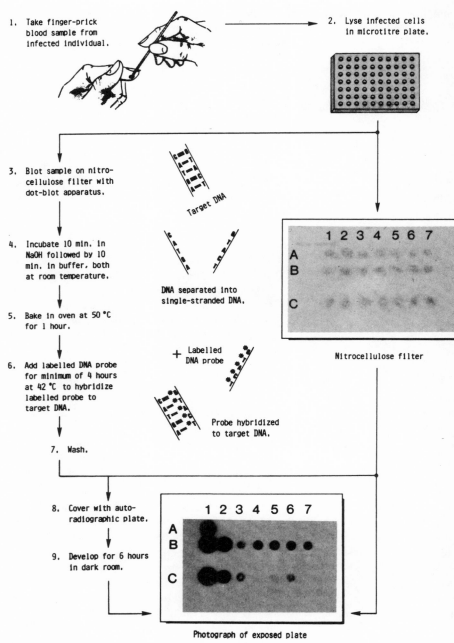

1. Take finger-prick blood sample from infected individual.

2. Lyse infected cells in microtitre plate.

3. Blot sample on nitrocellulose filter with dot-blot apparatus.

Target DNA

4. Incubate 10 min. in NaOH followed by 10 min. in buffer, both at room temperature.

DNA separated into single-stranded DNA.

5. Bake in oven at 50 °C for 1 hour.

6. Add labelled DNA probe for minimum of 4 hours at 42 °C to hybridize labelled probe to target DNA.

+ Labelled DNA probe

Probe hybridized to target DNA.

Nitrocellulose filter

7. Wash.

8. Cover with autoradiographic plate.

9. Develop for 6 hours in dark room.

Photograph of exposed plate

FIGURE A–2. The logic of DNA hybridization probes. From WHO 1986.

214

Various modifications of the basic pattern, such as "dot blots," have been developed for relatively rapid identification of complementary DNA sequences in many different kinds of biological materials.

A major difficulty with hybridization probes is their high cost, on the order of ten times higher than immunodiagnostic techniques using monoclonal antibodies, which are themselves rather expensive. Additional problems are the need for special equipment and the long time needed for the entire procedure including autoradiography, which alone requires at least several days, an ultralow temperature freezer ($-70°C$), and darkroom equipment for photographic development.

As mentioned previously, the need for radioactivity causes formidable problems in shipping, storing, using, and disposing of radioisotopes. For these reasons, nonradioactive labels are in great demand, but these often have high background, low sensitivity, and other technical drawbacks. In one promising method the DNA strands are conjugated to the vitamin biotin. After incubation of the denatured nucleic acid strands, hybridization is detected by the addition of streptavidin linked to an enzyme such as alkaline phosphatase or horseradish peroxidase, or to a fluorochrome. The conjugated streptavidin binds avidly to the biotin on the probe, and the combination is visualized by addition of an appropriate enzyme substrate, or with ultraviolet light, as appropriate. Another proposed method involves a detection reagent that incorporates an enzyme that catalyzes a chemical reaction that gives off light, which is detected visually or by photographic film. These methods are exactly analogous to those used for immunological localization. Chemical reactions are more rapid than autoradiography, and the reagents are more stable than the $^{32}P$ that has a half-life of only about two weeks. However, additional work is needed for chemical labels to achieve the sensitivity of $^{32}P$-labeled DNA.

*Molecular epidemiology.* DNA probes can find application in the emerging field of molecular epidemiology, which offers to investigators a higher "resolving power" for identification of organisms than was available heretofore. As an example, assume that three children from a village all have severe diarrhea, and that stool cultures from all three suggest that a species of *Shigella* is the causative agent. Conventional laboratory diagnosis by bacterial culture can group the organisms on the basis of their ability to ferment lactose or mannitol, presence or absence of ornithine decarboxylase, colony growth properties, and so on. There is substantial overlapping of serologic behavior among the different forms of *Shigella,* and some antigens are shared with other enteric bacilli.

Given the appropriate oligonucleotide probes, it may be possible to determine directly, with little or no need for culture, biochemical, or serologic tests, not only whether the organism is *Shigella dysenteriae, flexneri,* or *sonnei,* but the precise genetic markers present in each isolate. Appropriate DNA probes might tell, for example, that children one and three are infected with the identical genetic type of *Shigella,* indicating a common source, while the organism in child number two is genetically distinct and was probably acquired separately from the other two. In this way the specific pathways of infection can be traced with greater assurance and a higher degree of specificity than can be attained with conventional bacteriologic methods.

In another example, there are about two dozen known varieties (serotypes) of the *Legionella* bacterium that causes Legionnaire's disease, distinguishable by different immunological markers. Diagnosis of the organism by means of antibodies is impractical. However, DNA oligonucleotide probes exist that are complementary to regions of the bacterial genome that do not code for the different antigenic types, and that are common to all serotypes. Applying these probes in a single screen can capture all forms of *Legionella,* which can then be distinguished, if needed, by serologic or other DNA probes.

*Restriction fragment length polymorphism (RFLP).* In this procedure, the total DNA of an organism, or a large part of the DNA, is subjected to digestion by an enzyme called a restriction endonuclease. Each of the hundreds of known restriction endonucleases cleaves DNA at a particular point in the DNA sequence. For example, enzyme 1 may recognize only -GGGAT- sequences and cut at -GG^GAT-. Enzyme 2 may cut every -AGAACT- and leave -AGA separated from ACT-. After incubation with one of these enzymes, a long stretch of DNA would be divided into pieces, called restriction fragments, whose length depends on the number and location of the specific target sequences recognized by that enzyme. If the fragmented DNA is then placed on a gel in an electrophoresis cell and subjected to an electric current, the pieces will migrate in a characteristic pattern according to their molecular weight. The pattern will be different if the original DNA has changed in base sequence in a way that is recognized by the particular restriction enzymes being used. In the example above, a mutation in an original -GGGAT- sequence (e.g., to -GGCAT-) will result in nonrecognition by enzyme 1 and leave the sequence intact. The resultant DNA fragments will have a different size distibution from the original form. Consequently there will be a different pattern of restriction fragment lengths on the gel. An analogous situation would apply for enzyme 2.

Analysis of natural populations of many kinds of bacteria, such as *Mycobacterium tuberculosis,* reveals a large number of different RFLP patterns that are stable over time. Clusters of identical patterns in infected individuals may indicate a common source of infection.

## Nucleic Acid Amplification

A major problem with DNA hybridization is that although it is highly specific it is relatively insensitive. A large number of copies of the target DNA sequence may be required: tens of thousands for ordinary signal detection by radioisotopically labeled DNA, and perhaps tens of millions for colorimetric labels. Where few organisms exist in a sample, their DNA may be too scarce to be detected by these means. If this nucleic acid could be chemically amplified, a large amount of DNA would be available for hybridization or other manipulation. Several methods have recently become available to amplify DNA a million-fold within a few hours. All are based on polymerase enzymes that act to assemble nucleic acids. The concept is simple: double-stranded DNA is denatured to separate the two single strands, each of which is then used as a template to make a complementary strand which binds to it, thereby making twice the amount of DNA originally present. The process can be repeated through many cycles, each of which doubles the number of specific DNA segments present, leading to a very large amount of identical material after several dozen cycles. The basic difficulty with this process is that the denaturation requires heat, which destroys the activity of most polymerase enzymes. Therefore in early versions of amplification reactions, fresh enzyme had to be added at each cycle. Subsequently a thermostable enzyme was discovered, as described in the next paragraph.

*The polymerase chain reaction (PCR).* This is the earliest technique to amplify DNA sequences. It was invented in 1983 and remains the dominant method, although non-thermal techniques have been described. PCR depends on a DNA polymerase enzyme extracted from *Thermus aquaticus,* a bacterium that lives in natural hot waters. This enzyme, known as *Taq* or *Taq polymerase,* survives elevated temperatures and has the ability to catalyze the duplication of fragments of DNA. The PCR process, patented by the Cetus Corporation, utilizes a thermal cycling machine that operates automatically. Several other amplification systems, based on generally similar principles, are available, and many companies produce the various reagents and machines required for DNA amplification. These methods have become extremely widespread and the range of applications employing the detection of small amounts of DNA is essentially unlimited.

In a typical PCR procedure, DNA is first heated to separate the two complementary strands. Two short pieces of DNA, called primers, are then synthesized. These mark the ends of the DNA segment that is to be duplicated. One primer contains a DNA sequence from one strand, and the other contains a sequence from the complementary strand. When the primers are added to a tube containing the separated strands of the original DNA, each primer attaches to its complementary sequence—one on each strand. The Taq polymerase is then added. This enzyme controls the addition of nucleotides to the "inner" end of each primer until the intervening sequences are filled in in both directions. At this point the desired piece of DNA has been copied. The preparation is heated again to separate the strands, the process is repeated, and the target DNA sequence is once again doubled. After 30 cycles more than a billion copies should be present.

*Assessment of hybridization methods.* Nucleic acid hybridization can provide highly specific identification of organisms directly at the genetic level. Conventional methods such as microscopy, cultivation in media, or animal inoculation are rendered unnecessary. Complex transport media and other means of keeping the delicate organisms alive are avoided because the nucleic acids are chemically stable and remain intact even after the organisms have died. Therefore hybridization procedures may work even in situations where *in vitro* culture would be impossible. For example, pieces of tissue fixed and embedded in paraffin blocks can be examined by PCR for decades after the original biopsy was made. The method can be applied not only to patient-derived specimens but also to insect vectors, foods, or other materials that may harbor the pathogen (or more specifically, the DNA sequence) in question.

Nucleic acid amplification through the polymerase chain reaction has become a major diagnostic technology in the United States and other wealthy countries. It remains to be seen whether it is reasonable or practical to apply PCR to the detection of infectious disease agents in developing countries.

Although PCR and DNA hybridization are conceptually straightforward, the procedures are very costly, technically complex, and require a well-equipped laboratory and highly trained technicians.

# APPENDIX B

# An Introduction to the Vaccines

## Vaccines

An excellent, well-documented review of the history of vaccines was given by Sasson (1988). Briefer historical summaries of vaccinology were prepared by Plotkin (1988) and Hilleman (1986a,b), who denoted the major "eras" shown in Table B–1. Robbins (1990) has described how

> interest in vaccines peaked in the 1930s and then declined with the introduction of antibiotics; physicians, public health officials, and manufacturers then put all their hopes on treatment. Throughout the world, vaccine research, development, and production contracted. . . . When the new molecular biology of the 1970s and 1980s opened the way to rapid progress, there were few institutions and researchers with the knowledge to exploit the new opportunities. Today, the scientific revolution in modern biology . . . is finally generating widespread interest in vaccine research.

The main types of vaccines, with examples of each, are listed and described below. Current preparations are by no means perfect. Various authors, such as McIntosh (1989), have described common problems and feasible approaches to improving vaccines intended for the EPI.

### Classification of Vaccines

*Conventional vaccines.* The vaccines in routine use are generally derived from organisms grown in culture medium, in embryonated eggs, or in tissue cultures. Despite great efforts at quality control and uniformity, such biological products are always subject to some variation. Organisms may change or mutate in culture, and unexpected properties may appear. Particularly where living cells or tissues are used to grow the organism, the possibility exists of undetected contaminants

TABLE B–1.   Eras in Vaccinology

| Period | Technology | Important Vaccines |
|---|---|---|
| 1789–1949 | Organs of animals<br>Embryonated hens' eggs<br>Whole bacteria<br>Bacteria toxoids | Smallpox<br>Rabies<br>Yellow fever<br>Typhus<br>Encephalitis[1]<br>Influenza<br>Cholera<br>Typhoid<br>Tuberculosis<br>Diphtheria<br>Pertussis<br>Tetanus |
| 1950–1980 | Cultivated animal cells<br>Bacterial polysaccharides<br>Extracted subunit antigens | Poliomyelitis<br>Adenovirus<br>Measles<br>Mumps<br>Rubella<br>Improved rabies<br>Pneumococcus<br>Meningococcus |
| 1981– | Subunit virus<br>Directed genetic change<br>Recombinant antigen production<br>Chemical synthesis<br>Identification of epitopes by<br>  monoclonal analysis<br>Engineered "polytopic" vaccines | Hepatitis B |

Source: Hilleman 1986b. Copyright © The Medical Journal of Australia 1986; 144:360–364. Reprinted with permission.

[1] Japanese B and Russian Spring Summer.

such as viruses with unknown consequences to persons who receive the vaccine (see chapter 5).

1. *The actual infectious agent.* As mentioned in chapter 1, in times past, smallpox pustules were powdered and inhaled in the risky procedure of variolation. That method often induced immunity but sometimes led to a fatal case of smallpox. Similarly, scrapings from "Oriental sores" (lesions of cutaneous leishmaniasis) were inoculated by ancient peoples of the Middle East onto the buttocks of small children to prevent the possible appearance of disfiguring facial lesions later in life.

2. *Closely related organisms stimulating cross-immunity.* Inoculation with cowpox to prevent smallpox by Jenner in the eighteenth century marked the beginning of modern vaccination (*vaca* means cow). Some recent authors refer to a "Jennerian approach" in immunization, for example when bovine or simian strains of rotaviruses are used as a vaccine to protect children against pathogenic human strains (Kapikian et al. 1986). Among bacteria, living BCG bacilli of several strains are given to induce protection against tuberculosis and, to some extent, against leprosy.

3. *Living attenuated organisms.* Cultured organisms are weakened in various ways so that they will remain infectious and stimulate immunity but not disease. The biological nature of attenuation, its history, and the characteristics of many attenuated organisms were well described by Almond and Cann (1984). Some vaccines in common use that contain attenuated organisms are listed in Table B–2. Among the problems with this familiar approach are (1) assuring that the attenuation procedures have really made organisms nonpathogenic while maintaining antigenicity; (2) the chance for reversion to virulence; (3) the possible introduction of genomic segments from the living pathogen into the vaccinated hosts; (4) the small but measurable incidence of severe adverse effects such as encephalitis; and (5) the need to maintain continuous viability of the organisms, usually through a cold chain, from the point of manufacture to the point of use.

There is no theoretical reason why this general approach would not work with larger organisms, such as parasites. Some living attenuated roundworm larvae have found limited commercial use in veterinary practice, and field trials have been carried out in cattle exposed to irradiated larval schistosomes.

4. *Killed whole organisms.* Heat, chemicals (formalin), or other methods are employed in production of inactivated (Salk) poliovirus vaccine, as well as typhoid, whole cell pertussis, plague, spotted fever, and other vaccines.

5. *Bacterial products.* The toxin released by the bacterial agents of tetanus or diphtheria may be altered chemically to a toxoid, which lacks the harmful properties but remains immunogenic. Some pertussis toxoid vaccines are in use.

6. *Subunit vaccines.* Constituent parts of some pathogens may be separated from the rest of the organism and prepared as a vaccine. The purified capsule polysaccharides of pneumococci or meningococci are good examples.

*Synthetic vaccines.* Much current work is directed toward the development of vaccines consisting of uniquely specific antigenic molecules of known structure. It is hoped that such materials, of uniform reagent quality, can be synthesized in production volumes, providing a standardized, inexpensive, and effective product.

TABLE B–2. Properties of Some Vaccines

| Vaccine | Effectiveness (%) | Duration of Protection | Fatality Rate | Cost |
|---|---|---|---|---|
| *Smallpox | >90 | >3 years | 1/200,000 | + |
| *Yellow fever | >90 | >10 years | $1/10^7$ | + |
| *Poliomyelitis | >90 | lifetime | $1/3 \times 10^6$ | + |
| *Measles | >90 | lifetime | $\sim 1/2 \times 10^6$ | + + |
| *Rubella | >90 | >10 years | $1/10^7$ | + + |
| Influenza | 70 | 1 year | very low | + + |
| Rabies | >90 | | 1/25,000 | + + + |
| Hepatitis B | >90 | 5 years | very low | + + + |

*Source:* Ada 1986.

*Live, attenuated agents.

1. *Chemically synthesized antigens.* If a specifically defined protein or polypeptide (amino acid chain) is determined to induce protective immunity, it may be possible to devise a total chemical synthesis to produce the antigen in pure form (Klipstein et al. 1986). A malarial sporozoite surface antigen vaccine has been made in this way.
2. *Recombinant DNA procedures* (see also Table B–3).

   a. *Antigens harvested from fermentation.* Standard genetic engineering practice for production of an antigen calls for the identification and excision of the gene responsible for its production, and insertion of that gene into a foreign (host) organism for expression. Such foreign gene products expressed in *E. coli,* yeasts, or animal cells may be harvested, purified, and made into vaccines.

   b. *Engineered living carriers.* Rather than collecting the foreign gene product from a fermentation vessel, it may be possible to utilize the host organism itself, complete with the new insert(s), as a live vaccine. Among the organisms proposed for this purpose is the classical vaccinia virus. Vaccinia is a feasible vector for new multi-insert vaccines because:

      1. It replicates well in a variety of mammals, including humans, but secondary transmission from person to person is rare. Note, however, the report of Baumgaertner et al. (1985) of a case in which vaccinia from a recent vaccination was transmitted to an unrelated person.
      2. Its genome contains a 28-kilobase sequence that is not necessary for successful viral replication and may be substituted by one or several antigenic gene sequences.
      3. Vaccinia vaccines used successfully in the past were easily prepared, stable, convenient, and provided long-term immunity.

*Antigens already tested in vaccinia vectors.* Several inserts have already been put into vaccinia vectors and tested in animals and humans. Those genes intended for human use include influenza virus hemagglutinin, and antigens of hepatitis B, herpes simplex, rabies, Epstein-Barr, and human immunodeficiency viruses. Veterinary vaccines include vesicular stomatitis, transmissible gastroenteritis and feline leukemia viruses; and malaria. All elicited antibody responses after a single dose, with the exception of hepatitis B antigen given to chimpanzees. In that case a second dose did produce a secondary response, and the animals were protected from clinical disease but not from infection. Several investigators have succeeded in inserting several of these genes into the large genome of the vaccinia virus.

TABLE B–3.    Recombinant DNA Approaches to Vaccines

---

*Subunit vaccines*
1. Expression of immunogenic polypeptides in genetically engineered prokaryotic or eukaryotic hosts
2. Carrier-derived synthetic peptides where amino acid sequences have been derived from cloned DNA

*Live vaccines*
1. Genetic reassortment *in vivo,* e.g., between the RNAs of different influenza viruses
2. Modification of DNA by *in vitro* mutation, e.g., the modification of cloned poliovirus infectious DNA
3. Modification of virus DNA *in vitro* using restriction enzymes, e.g., adenoviruses, herpesviruses
4. Antigens cloned in vaccinia virus, e.g., hepatitis B surface antigen, influenza HA

---

*Source:* Harris 1984.

*Other engineered living carriers.* Among other potential vectors that have been suggested as potential vaccines are herpes simplex virus, 17D yellow fever virus, poxoviruses, adenovirus, *Shigella, Salmonella typhimurium,* and *Escherichia coli.*

Genes coding for antigenic regions of enteric pathogens may be packaged into avirulent *E. coli,* which multiply in the gut and may be expected to induce immune responses against a variety of infectious agents.

A strategy employing living avian pox vectors, incapable of replicating in mammals, has been described by Baxby and Paoletti (1992). Fowlpox viruses constructed to express rabies glycoprotein were inoculated into mice, cats, and dogs, all of which seroconverted to an extrinsic rabies immunogen. Similarly, canary pox viruses with a recombinant rabies glycoprotein insert have produced strong specific antibody responses in chimpanzees. Avian pox viruses have been promoted for human use. Because they are nonreplicating they should not disseminate within the vaccinee, from a vaccinated to a nonvaccinated individual, or contaminate the environment.

*Concerns about recombinant viruses and bacteria.* Viral vaccines can be frozen and their genomes stabilized. Bacterial plasmids are more unstable, and selection by antibiotic resistance is necessary to pick out strains carrying the desired antigenic capabilities. Because many bacteria multiply only about ten times before they are eliminated from the host, reversion to virulence in the vaccinated host seems unlikely.

Vigilance is always needed to monitor and control the level of attenuation of microbial vectors. They must be capable of replicating, and of expressing the inserted foreign genes, which should not be lost or discarded. They must not revert to a harmful wild type. The act of inserting a foreign genomic segment may increase or decrease their underlying natural virulence. Conceivably, changes in host range or tissue tropism could result from genetic insertions or from the expression of different envelope characters by a recombinant virus.

> When you change a virus you don't always make it better. We can't assume that vaccinia is going to behave as vaccinia just because it is mostly vaccinia. The great likelihood is that it is going to remain the same, or be better, but we have to be aware of the possibility that it may develop new properties. Studies should be done on individual recombinant viruses as they are developed, to address this concern (Quinnan 1985).

Many investigators are concerned about the effects of infection with novel living agents such as recombinant vaccinia virus on persons with immunodeficiency states. This concern has become more acute after the deaths of two volunteers with AIDS who had been inoculated with lymphocytes previously infected with a recombinant vaccinia virus. Guillaume et al. (1991), who reported these cases, emphasized the risk of vaccinia disease in mass-vaccination programs that use recombinant vaccinia virus expressing HIV genes, especially in Africa "where exclusion of seropositive patients before vaccination would cause insurmountable logistic difficulties and costs."

## Anti-idiotypic Antibodies as Vaccines

According to the old lock-and-key theory of antibody function, the physical shape of an antibody reflects a sort of inverse pattern of the evoking antigen. It was discovered in the early 1970s that antibodies incorporate regions that are themselves antigenic and can stimulate the production of other antibodies, which might become part of a regulatory feedback loop controlling the production of immunoglobulins. A second antibody directed against the first would then include a structure ("internal image") similar or identical to the original antigen; this has been termed "antigenic mimicry." Such second antibodies, termed anti-idiotypic antibodies, anti-idiotypes, or simply anti-Id, might then be used as a sort of vaccine incorporating a surrogate antigen. They have been found to be protective against certain infections in experimental animals.

> The advantages of using anti-Id are considerable. As for subunit vaccines produced by gene cloning or peptide vaccines by chemical synthesis, the anti-Id approach would overcome the limitation of antigen source and the inherent hazards associated with other forms of vaccines. An additional advantage would be the potential elimination of defective configurational problems which beset antigens produced by gene cloning or synthetic peptides. This approach, when successful, would also be monospecific in that it provides a means of immunizing against single antigenic determinants, uncomplicated by associated and frequently undesirable side reactions. The major disadvantage of an anti-Id vaccine would be its inherent uniqueness which could make it potentially impractical to be applied to the polymorphic human population (Liew, 1989).

Applications of anti-idiotypic antibodies in humans have been limited to cancer patients, with results suggestive of further investigation. There has been little or no application thus far to the great endemic diseases of developing countries.

## Microencapsulation

One technique for introducing vaccines is microencapsulation.

Microencapsulation of vaccines into copolymer microspheres results in a dry powder-like material with extended shelf life. The polymer coats are biodegradable after specific time intervals within the host. Mixtures of microcapsules with different polymer characteristics can be administered so that the contained vaccines can be "pulse-released" at various times, providing

> a vaccine delivery system in which a primary and any number of booster vaccine releases can be programmed over a two-year period, to match current vaccination schedules with a single administration. Second, the functional independence of the components in a microsphere mixture indicates that different vaccines can each be microencapsulated to release by their own unique pulsed schedule which can be administered as a single mixture without the components affecting each other (Eldridge et al. 1991).

## Effect of Differing Routes of Administration

Living recombinant vaccines may show different immunogenicity or efficacy depending on the route of administration. For instance, *Vaccinia* virus without inserts is virulent by intraperitoneal injection but not by other routes. After insertion of antigen-coding regions for herpes simplex or hepatitis B, the recombinant *Vaccinia* virus lost its virulence even when injected intraperitoneally (WHO 1985).

It is uncertain whether multiple inserts will provoke differential responses that vary among individuals or groups. Live or killed conventional vaccines contain hundreds of antigens. In multiple-insert living vaccines it remains to be determined whether the various foreign genomic segments and their evoked antigens will act cooperatively with a synergistic effect, or antagonistically to reduce the immune response to individual elements.

In a novel procedure Tang et al. (1992) inoculated mice via the ear skin with gold microprojectiles coated with a plasmid expressing the gene for human growth hormone or for human alpha-antitrypsin. When the gold microprojectiles were injected by syringe needle there was no response, but when a hand-held "biolistic" device was used to propel the DNA directly into cells antibodies were produced in mice of an appropriate strain. In this system the need for protein isolation and purification, and for adjuvant, preservative, or other components of a complete vaccine, is eliminated. It remains to be seen whether this procedure will be usable in humans.

## Combinations of Vaccines Administered Together

Numerous fixed vaccine combinations have been proposed, and several, such as DPT (diphtheria toxoid + whole-cell pertussis + tetanus toxoid) and MMR (live attenuated measles + mumps + rubella), are in everyday use on a global basis. In addition, different vaccines are often used simultaneously without adverse effects, even in newborns. In areas endemic for poliomyelitis, the WHO recommends immunization at birth with both the live bacterial vaccine, BCG, and with the first dose of trivalent oral polio vaccine.

Considering only live viral vaccines, Hilleman et al. (1973) listed three characteristics required for effective combinations: (1) there must be no interference between viruses in the combined vaccine; (2) the duration of immunity must be as long for each separate virus in the combination as it would be if the vaccines were administered separately; and (3) clinical reactions to the vaccine should not be greater than would be expected by the administration of the most reactive component as a single vaccine. The outstanding example of a live virus vaccine combination that adheres to these criteria and provides long-lasting clinical immunity is MMR.

A potential problem with combination vaccines is the differential loss of potency known to occur among the individual components. Consequently, it may be difficult to test independently for each element of the mixture and to decide what to do with the vaccine if one of several components has lowered potency.

*Advantages.* Vaccine combinations can lead to streamlining of immunization activities. In addition to biological simplicity, a combined vaccine reduces the number of contacts with the health system, which can increase cost-effectiveness and allow better coverage for hard-to-reach groups. Because they provide a way to maximize coverage, vaccine combinations support the policies of the Expanded Programme of Immunization and of the Children's Vaccine Initiative.

Cohen et al. (1973) discussed the use of a quadruple DPT + polio vaccine in the Netherlands, reporting that no cases of polio, tetanus, or diphtheria had been reported in infants receiving at least three doses of the vaccine since this combination was first used in 1962. Pertussis figures were not available. The vaccine was administered on a four-dose schedule at 3, 4, 5, and 12 months of age, with subsequent DT + polio boosters. Additional advantages of the combined administration were:

1. Provided that health authorities are willing to establish and stick to a consistent schedule, the immunization becomes a routine part of health care.
2. The simplicity of administration makes the program easily understood for both medical personnel and parents.
3. Assuming that the vaccine combination has minimal side effects and provides good clinical protection, simultaneous administration can offset the apathetic attitude toward immunization that arises when population incidence decreases. If a disease drops to a low incidence and the vaccine against that disease is administered singly, people may become less inclined to seek immunization since the disease no longer poses an obvious threat. If, however, the vaccine for a relatively uncommon disease is administered as part of a routinized package, especially when some other vaccines in the combination protect against readily visible or prevalent diseases, high immunization coverage, and protection of the population, will continue against the low-incidence disease.

*Concerns and disadvantages.* There are two main concerns when vaccines are combined: possible enhancement of adverse effects, and possible interference with immune response, particularly in live viral vaccines.

Induced antibody titers may vary in predictable ways when vaccines are administered in certain combinations. For example, Felsenfeld et al. (1973) found that the then-prevalent cholera and yellow fever vaccines, administered simultaneously (or within three weeks of each other), reduced circulating levels of antibody to both vaccines. The authors tested serum samples from individuals who had been vaccinated against smallpox, cholera, or yellow fever, as well as from those who had received yellow fever and cholera, smallpox and cholera, or all three vaccines. When cholera vaccine was given three weeks or less before yellow fever vaccine, the cholera antibodies subsided at a "natural" rate, but yellow fever antibody titers were lower than when only yellow fever vaccine was used. When yellow fever vaccine was given before cholera vaccine, post-cholera vaccine titers were lowest when the interval between vaccines was greater than three weeks; yellow fever titers were again lower than those found for yellow fever vaccination only. Although circulating titers are not necessarily representative of protection, possible antagonism is an important consideration in the planning of cholera campaigns in areas of high yellow fever incidence.

Interactions may be subtle. Kaplan et al. (1984) reported that seroconversion to oral polio type 1 vaccine (but not types 2 and 3) was reduced when Peace Corps volunteers also received simultaneous cholera vaccine. Similarly, seroconversion to an attenuated rotavirus vaccine (RIT 4237) was inhibited when it was administered together with bivalent oral polio vaccine, whereas responses to the polio vaccine did not differ whether given alone or with the rotavirus vaccine (Giammanco et al. 1988). Although most combinations of multiple live vaccines have shown no interference, there is no guarantee that each component will function independently of all others.

An exhaustive review of vaccine-vaccine, vaccine-immunoglobulin, and vaccine-drug interactions was published by Grabenstein (1990), who made the point that sometimes there may be a specific reason to put the immunologic brakes on a live virus vaccine:

> Historically, it is interesting to note that an early attenuated strain of measles vaccine (Enders' original Edmonston B strain) was intentionally administered with immune globulin or hyperimmune measles immune globulin to reduce that vaccine's adverse-effect profile and restrain the replication of the insufficiently attenuated vaccine formulation (Grabenstein, 1990).

## Improving the Immunogenicity of Vaccines

It has been observed repeatedly that simple antigens, such as peptide fragments produced by chemical or recombinant means, are poorly antigenic. Numerous attempts are being made to enhance the immunogenicity of simple antigens and of the vaccines that contain them. Among the ways to do this are:

1. Coupling short peptide antigens to larger carrier proteins such as diphtheria toxoid
2. Concentrating short peptide antigens into aggregates such as liposomes or micelles
3. Incorporating adjuvants into the vaccine

The potential impact of microencapsulation, described earlier, remains to be determined.

## Adjuvants

Adjuvants are immunopotentiating agents added to specific antigens to enhance appropriate immune responses of vaccinees. For novel vaccines built around specific antigens produced by chemical synthesis or by recombinant DNA methods, some sort of adjuvant is usually necessary because the isolated antigens are not nearly so immunogenic as those in the complete organism. Adjuvants can also influence the immunoglobulin class of antibodies produced by the host.

The classical water and oil emulsions such as Freund's complete or incomplete adjuvants commonly used in experimental laboratory animals may cause abscess and granuloma formation and cannot be considered for use in humans. Other adjuvants include various emulsions; saponin; liposomes; entire bacteria such as BCG or bacterial cell wall components including muramyl dipeptide (MDP); and "mineral gels" such as aluminum hydroxide. Toxoids precipitated with aluminum hydroxide or aluminum phosphate are more immunogenic than soluble toxoids. When used as coupling agents, toxoids are themselves adjuvants. Monoclonal IgM in a mouse system also displays adjuvanticity (Harte et al. 1983). Interleukins and other cytokines have adjuvant activity, but no commercial use of such products has been approved in humans. At present, mineral gels are the only adjuvants approved for incorporation into vaccines for humans (Bomford 1989).

Several workers have described immunostimulating complexes (ISCOMs) that consist of antigens (generally viral envelope glycoproteins) attached to multimeric physical structures about 40 nanometers in size. These include a built-in adjuvant, the glycoside Quil A. ISCOMs are said to be 10-fold more immunogenic than the same amount of antigen in other forms (Morein 1990), but they may also induce hypersensitivity reactions rather than protection.

*Mechanisms for adjuvanticity.* Bomford (1984) has pointed out some of the difficulties in understanding mechanisms of action of immunological adjuvants:

1. Many adjuvants are complex and provoke a variety of responses when administered, some irrelevant to the adjuvant effect.
2. Because the immune response is a multistep process and occurs through a variety of pathways, adjuvants can theoretically act on a variety of cells and in a variety of steps in each pathway. As the processes of immune response are not yet fully elaborated, our understanding of how adjuvants influence immune response is similarly unclear.

Several mechanisms for adjuvant action have been proposed, including slow release of antigen, uptake by macrophages or other effector cells, lymphocyte activation, and stimulation of complement or lymphokine synthesis. These are too complex to be discussed here. The biomedical innovator interested in producing vaccines against parasitic diseases should heed Bomford's (1989) comment that stimulation of cell-mediated immunity against parasite antigens is precisely the mechanism by which the infections themselves induce major pathology. Therefore the adjuvant portion of a vaccine may be more difficult to develop than the antigenic elements. In addition, there are significant differences in adjuvant effect from species to species and among different individuals within species. The extent of this genetic variability is not yet known in human populations, but it may be significant.

# APPENDIX C

# An Introduction to the Biology

Foremost among the elements that determine whether a biomedical innovation will serve its function within any particular environment is a good understanding of the basic biology of the human hosts, the pathogens, the vectors (if any), and the interactions among them. It is clearly impossible to present in a few pages all of the factors that must be taken into account. Some have been mentioned in previous chapters. Those considered most significant are discussed briefly here.

## Host Considerations

### Immunity

The major problem facing vaccine development is how to identify the right kind of long-lasting immune response. This is the area of vaccine production and use in which the majority of unidentified and least understood problems of immunology remain. . . . The importance of understanding the nature of the host's response is demonstrated by the knowledge that the wrong or an inadequate immune response may lead to pathology or immunosuppression, not protection (Ogilvie 1988).

"Immunity" is often considered equivalent to "resistance," embracing all the characteristics of an individual that confer defense against a specific agent at a particular point in time. However, people exist not solely at one of the two poles of solid resistance or complete susceptibility, but in a continuous spectrum of gradations in between. Resistance to a particular infectious agent may be inborn. Two examples: only a few species of mammals can be infected with leprosy bacilli, and most people of African origin are innately incapable of becoming ill with vivax malaria. On the other hand, originally susceptible individuals may acquire immunity by effective contact with a particular infectious agent or with its products. The active immunity so stimulated usually takes some weeks to develop and may thereafter be essentially lifelong, as with yellow fever or smallpox immunization, or may wane after a variable period.

Immunity may also be acquired passively, as in the transfer of blood products containing antibodies. Newborns acquire antibodies of the IgG class that pass through the placenta from their mothers, and also receive some protection through immunoglobulin present in breast milk. Deposit injections of immune serum globulin (ISG or gamma globulin) are sometimes given to Western travelers to the tropics. The protection afforded by passive immunity is relatively short-lived, depending primarily on the amount of protective material transferred and the body size of the recipient. The transferred antibodies diminish and eventually disappear.

The use of monoclonal antibodies has been suggested to confer passive immunity to challenge. Their possible utility in developing "vaccines" based on anti-idiotypic antibodies, mentioned in Appendix A, remains to be evaluated. Such application might be feasible where a protective antigen is in short supply and cannot be synthesized by chemical or recombinant DNA techniques. However, most available monoclonal antibodies are derived from mice and cannot be injected with impunity into humans.

### Constituents of the Immune Response in Humans

Responses to an invading pathogen or a nonliving antigen may be specific or nonspecific. The nonspecific part is typical of generalized inflammation and may involve secretion of hormone-like messenger molecules that stimulate responses such as the activation of white blood cells. Lymphokines, cytokines, interleukins, interferons, and tumor necrosis factor are examples of nonspecific humoral messengers. The specific responses are generally divided into humoral and cellular "arms." This is an artificial distinction because the specific humoral effectors, the immunoglobulins, are produced by B lymphocytes ("B-cells"), which are themselves controlled by other soluble chemicals in the blood. The main effector cells of immunity are T lymphocytes. Stimulation of lymphocytes involves very complex interactions between lymphocyte surface proteins and soluble factors, and details should be sought in a current textbook of review of immunology.

The important implications for persons interested in vaccines are (1) lymphocytes of each individual have inherent limitations in the range of antigens that they can recognize, and of antibodies that they can synthesize; (2) the functioning of effector cells depends on intracellular processing of proteins, so that vaccine components must be able to enter the appropriate cells in which they can begin to generate the immune response; and (3) the mechanisms that protect against different pathogens (or against separate life cycle stages of the same organism) may be entirely different. For example, antitoxins may counteract toxins produced by bacteria, but the bacteria are not killed. Antibacterial vaccines may induce host reactions that cause clumping or lysis of the bacteria, or facilitate their uptake by host phagocytic white blood cells.

An interesting interplay among antigen type and host capability is found in vaccines against the bacteria that produce severe respiratory infections and meningitis in many parts of the world. Examples of such bacteria are *Streptococcus pneumoniae, Haemophilus influenzae* type b (Hib), and *Neisseria meningitidis.* All are surrounded by polysaccharide-containing capsules that hamper destruction by host cells. Capsular polysaccharides are T-cell independent (TI) antigens. They can activate B-cells directly with minimal participation of T-cells. In TI responses, only certain subtypes of antibodies are produced and memory cells are not generated. In contrast, T-cell dependent (TD) antigens induce all kinds of antibodies and generate memory cells, important in subsequent infections. In general, children under two years of age and elderly people, high risk groups for infections with these pathogens, react poorly to TI antigens, whereas TD antibody responses can be obtained from birth onwards. When bacterial polysaccharides are used alone as vaccines, they induce TI responses, but when conjugated to proteins such as tetanus toxoid, they become TD antigens and can be used to protect infants and children under two.

The route and manner of immunization are associated in a complex way with the types of immune system responses that are elicited. When most people think of immunization they probably have in mind the traditional ''jab'' or syringe-needle method. However, the original vaccination (in the strict sense of the word, using vaccinia virus against smallpox) was done routinely by scratching or abrading the skin and rubbing in the vaccine.

Many of the diseases for which vaccines are used or considered affect the gastrointestinal, respiratory, or genital tracts, all of which are lined with mucosal membranes. The presence of a common mucosal immune system is now recognized, in which antigen-sensitized cells from the underlying lymphoid tissue migrate through the body to other mucosal sites, including the mammary glands. Oral or intranasal immunization has much to recommend it for combatting pathogens that enter through mucosal membranes, particularly of the gut or respiratory system. The intention of the Children's Vaccine Initiative and other programs to develop vaccines suitable for oral administration brings the mucosal immune system into special prominence for vaccine designers (McGhee et al. 1992).

For invasive interventions such as immunizations it is essential to know whether the intended subjects will react as expected to the vaccine. The obvious differences between healthy adult male volunteers in western countries and real populations in developing countries raise many questions, including:

1. Does nutritional status affect immunologic responses?
2. How do past or current infections with various agents affect host response?
3. How do age and sex influence immune protection?
4. Do genetic differences among different ethnic groups affect the efficacy of these interventions?
5. Do different human behaviors, such as breast- or bottle-feeding, make any difference?

## Nutrition

The relationship between nutrition and immunity has been explored many times, with reviews by Keusch (1984, 1991), Chandra (1988, 1991) and others. Since socioeconomic factors also correlate with the degree of malnutrition, a precise direct relationship between nutrition and immune responsiveness is difficult to establish.

*Cell-mediated responses.* The primary effect of malnutrition on the immune system is a depression in cell-mediated responses such as delayed hypersensitivity. Of the two major lymphocyte classes, T-cells are more affected than are B-cells, as shown by the usually adequate levels of immunoglobulins in malnourished individuals. Chandra (1991) reports a sharp reduction in proportion and absolute number of T-lymphocytes in protein-energy malnutrition (PEM). Altered T-cell regulation of IgE production in PEM can lead to unusually high levels of IgE and can contribute to IgE-mediated responses to certain viral pathogens (Keusch 1991). Other adverse immunologic effects of PEM are: thymic atrophy; decreased spleen weight; impaired lymphokine production; and anergy in delayed cutaneous hypersensitivity (Delafuente 1991). Katz (1983) focused on the relationship between marasmus or kwashiorkor and immunity and found depression of cell-mediated immunity. Diseases usually countered by cell-mediated immunity (such as measles, herpes simplex, tuberculosis, and candidiasis) are more severe in malnourished children, particularly in those with edema. In a study by McMurray et al. (1979), depressed cellular responses were noted in a group of malnourished 8-week-old Colombian infants who had been given BCG at birth. For moderately malnourished children, depressed immunity was still apparent at 2 years of age, as measured by delayed hypersensitivity to the PPD test. Moderate nutritional deficiencies in early development can cause lymphoid maldevelopment.

*Immunoglobulin production.* Reports are contradictory regarding the serologic response to immunization in cohorts of malnourished children. Most investigators find no significant differences in blood serum titers between malnourished and properly nourished vaccinees (Table C–1). Greenwood et al. (1986) closely followed young children in Nigeria given a full course of EPI immunizations. No significant correlations were found between any anthropometric measurements at

TABLE C–1.    Antibody Response to Measles Vaccine, by Nutritional Status, Hyderabad, India, 1984

| Nutritional Status[a] | Nutritional Grade | Number of Children | Percent Having Protective Titer |
|---|---|---|---|
| >90 | Normal | 6 | 83.5 |
| 76–90 | Grade 1 malnutrition | 44 | 81.8 |
| 60–75 | Grade 2 malnutrition | 52 | 92.3 |
| >60 | Grade 3 malnutrition | 14 | 85.7 |
| Total | | 116 | 87.1 |

*Source:* Adapted from Bhaskaram et al. 1986.

[a] Percent of standard weight/age

the time of immunization and antibody responses to DPT, polio, measles, meningococcal, or typhoid vaccines, nor between nutritional state and response to BCG. Some researchers have commented on a paradoxical but nonspecific increase in IgG level that often accompanies malnutrition. Some studies find depressed responsiveness to live poliovirus (especially types II and III), measles, or other antigens. The older literature showed some evidence that administration of BCG failed to produce subsequent tuberculin activity, although the situation is clouded by the fact that BCG is usually administered to newborns. Hafez et al. (1977) studied humoral immune responses to poliomyelitis, measles, and diphtheria vaccines in Egyptian children presenting with marasmus or kwashiorkor. Despite elevated total serum immunoglobulin levels, specific antibody responses were inhibited. Those children with kwashiorkor showed reduced levels of all complement components, indicating impaired protein synthesis in liver, and children with marasmus showed reduced levels of C3.

The production of secretory IgA does seem to be impaired in malnutrition. For example, Chandra (1975) compared 20 normal and 20 malnourished Indian boys 1 to 4 years of age, immunized with one dose of measles or poliovirus vaccine. Seroconversion and serum IgA levels were comparable after vaccination, but secretory IgA was delayed and showed a lower titer for the malnourished group, significant at the .05 level. Since total protein and albumin concentrations were comparable in both groups, reduced secretory IgA response seemed selective, leading Chandra to conclude that immunity may well be reduced after measles infections or gastroenteritis. Reduced mucosal immunity may permit systemic spread of otherwise localized infections. Such systemic invasion would induce a lymphoid response and may be responsible for the high levels of serum antibodies seen in such cases. The slower recovery could increase the period of contagiousness and alter the epidemiology of the disease.

Interferon production may (in marasmus) or may not (in kwashiorkor) be affected by protein-energy malnutrition (PEM). In children with PEM, increased consumption and reduced synthesis of complement factors, in conjunction with reduced phagocytic activity, can lead to increased susceptibility to bacterial infections (Keusch 1984).

*Anemia.* Bagchi et al. (1980) concluded in India that humoral response was not adversely affected in anemic subjects after immunization with diphtheria and tetanus toxoids. Preimmunization levels of serum IgG were in the normal range in 14 anemic individuals, defined as having a hemoglobin of less than 11g/dl, and in 24 normal control children aged 2 to 10 years. Two weeks after vaccination, serum IgG rose for both groups although T-lymphocyte levels were lower in the anemics, both initially and after immunization. There is no consensus of authors about the effect of iron deficiency on cell-mediated response.

Vitamin A, iron, zinc, and folic acid deficiencies may all compound the effects of infections. The role of iron appears to be particularly complex. Patients with iron-deficiency anemia tend to show high levels of transferrin, which inhibits bacterial growth in serum. Keusch (1984) noted that while a lack of iron leads to depressed bacteriocidal activity by white blood cells, doses of supplemental iron, coupled with low serum transferrin, may allow some microorganisms to proliferate. It has been found (Oppenheimer et al. 1986) that iron supplementation increases the clinical severity of malaria, suggesting that a moderate degree of iron deficiency may actually be useful in inhibiting malarial infection. In addition, calcium, magnesium, copper, selenium, and other trace elements may have significant effects on immunocompetence. Zinc deficiency has many adverse effects on immune function, including hypogammaglobulinemia, abnormal cell-mediated immunity with decreased mitogenic responses, decreased circulating T-lymphocytes, thymic involution, and decreased thymic hormones (Gershwin et al. 1991).

## Other Infections

*Malaria and other parasitic infections.* Malaria is endemic in many areas of the tropics and must be taken into account in any planned trial of a vaccine. Acute attacks of malaria were found by Greenwood and Whittle (1981a) to produce Wasserman antibodies similar to those found in syphilis infections, leading to false diagnoses of syphilis in tropical areas. Malarial attacks can also inhibit the response to tetanus, typhoid, and meningococcal vaccines (Greenwood and Whittle 1981b). In Northern Nigeria, Greenwood et al. (1986) found significant correlations between serum prealbumin levels and the response to group A meningococcal vaccine, and between albumin levels and response to group C meningococcal vaccine in malarious children, concluding that this infection may have a depressive effect both on serum prealbumin and albumin levels and on responsiveness to meningococcal polysaccharide vaccines. Other investigations have suggested that whereas subclinical malaria may impair the response to polysaccharide vaccines, it does not do so to measles or MMR vaccine. Paradoxically, malarial infections may also lead to production of excessive levels of IgG, which may not be specific to any particular antigen. Although most parasitic infections do not lead to diminished response to vaccines, African trypanosomiasis may cause both hypergammaglobulinemia and nonspecific immunosuppression (Katz 1983).

*Acute respiratory infections.* In developing countries these infections are extremely common in infants as a result of low birth weights, malnutrition including vitamin A deficiency, indoor air pollution, and more recently, HIV infection. Their influence, if any, on immune status remains unclear.

*Effects of immunization on nontarget infections.* Bhaskaram et al. (1986) found that cell-mediated responses were similar for malnourished and healthy children aged 9 months to 3 years in India (see also Table C–1), both before and after immunization, but morbidity resulting from other infections was less for vaccinated children, a difference attributed to the activation of nonspecific immune response mechanisms. It is possible also that other factors, such as differences in the interest and skill of the mothers in caring for their children's health, played a role in the reduced morbidity. Peltola et al. (1991) reported that oral B-subunit/whole-cell cholera vaccine given to Finnish travelers to Morocco induced 52% or more protection against diarrhea caused by enterotoxigenic *Escherichia coli* bacteria, in a Jennerian-type interaction.

The reported excess mortality from all causes following administration of high-dose Edmonston-Zagreb measles vaccine was discussed in chapter 5. If this phenomenon is real, no mechanism for its mode of action has been proposed.

## Immunization in the Immunocompromised Host

While there are no adverse reactions to inactivated viral or bacterial vaccines, many persons have expressed concern about the use of live vaccines in cases of natural, induced, or acquired immunodeficiency. Reports have varied. Lawrence et al. (1988) studied 346 children immunized with attenuated live varicella virus while in remission from acute lymphoblastic leukemia, concluding that the incidence of zoster was no greater than that following natural varicella infection. Various authors have reported live vaccine-associated poliomyelitis, measles pneumonia, and BCG dissemination in immunocompromised or immunosuppressed children. A case of paralysis in a non-HIV immunocompromised adult following oral polio vaccination in Canada was reported by Mathias and Routley (1985).

*Acquired immunodeficiency syndrome (AIDS).* The subject of AIDS in developing countries, particularly in Africa, has received an enormous amount of attention, and is changing so rapidly that it cannot easily be summarized. The number of studies and recommendations on immunization of HIV-infected children grows rapidly and continuously (Table C–2). For newborns, serologic testing for HIV antibodies reveals only the status of the mother because of the passive transfer of IgG.

TABLE C–2. Recommendations for Routine Immunization of Asymptomatic and Symptomatic HIV-Infected Children, World Health Organization and Immunization Practices Advisory Committee

| | WHO | | ACIP | |
|---|---|---|---|---|
| *Vaccine* | *Asymptomatic* | *Symptomatic* | *Asymptomatic* | *Symptomatic* |
| BCG | Yes[a] | No | No | No |
| DTP | Yes | Yes | Yes | Yes |
| OPV | Yes | Yes | No | No |
| IPV | Yes | Yes | Yes | Yes |
| Measles | Yes | Yes | Yes[b] | Yes[b] |
| Influenza | | | No | Yes |
| Pneumococcal | | | No | Yes |
| HbCV[c] | | | Yes | Yes |

*Source:* Onorato IM. Childhood immunization, vaccine-preventable diseases, and infection with HIV. *Pediatric Infectious Diseases J.* 6: 588–595, 1988. © Williams & Wilkins, 1988.

[a]Where the risk of tuberculosis is high, WHO recommends BCG at birth for asymptomatic HIV-infected children; where the risk of tuberculosis is low, BCG may be withheld.

[b]ACIP recommends that MMR be given routinely for asymptomatic children and considered for symptomatic children.

[c]*Haemophilus influenzae* type b conjugate vaccine.

The risk of accelerating HIV infections through immunization appears low compared to other natural risks. HIV infection leads to malabsorption, malnutrition, and weight loss, and the response to nutritional support is limited by severe systemic infection. HIV-infected children have markedly elevated serum immunoglobulins, particularly IgG (Onorato et al. 1988). Although the situation is complex, infants not severely ill with opportunistic infections generally have normal *in vitro* lymphocytic mitogenic responses. In such children specific antibody response to administered antigens is generally low, but may be sufficient to be protective. Surveys of immunogenicity of recommended vaccines have been limited to small numbers of children, with variable results. For example, four studies using tetanus toxoid showed 100%, 85%, 60%, and 40% responders (Onorato et al. 1988). Overall, although authorities disagree on some specifics, the greater risk of severe complications of infection with the pathogens of vaccine-preventable disease argue for normal immunization of HIV-infected children, with the exception of BCG in children with symptomatic immunodeficiency.

The proposed widespread use of genetically engineered living carriers, such as vaccinia virus, in developing countries must be tempered by consideration of the prevalence of HIV or other causes of immunodeficiency (Guillaume et al. 1991).

## Age: Infancy

The infant immune system must be considered as different from that of older children and adults. It has been mentioned that bacterial capsular polysaccharide antigens of some important pathogens such as *Streptococcus pneumoniae* and *Haemophilus influenzae* type b, agents of pneumonia, meningitis, and otitis media, are ineffective in infants, unless those polysaccharides are first converted to T-dependent antigens by conjugation to a protein carrier.

*Maternally-transferred antibodies.* Maternal IgG has a half-life of 3 to 4 weeks after birth (Insel 1988). As a practical consequence, there may be difficulty in immunizing small infants in whom maternally derived antibody levels are high. This issue is prominent for measles live-virus vaccine (Chapter 5), but may be significant also for nonliving antigens. For example, Booy et al. (1992) in the UK have compared antibody titers against combined diphtheria, tetanus, and pertussis, when given on an accelerated schedule 2, 3, and 4 months of age versus the conventional schedule of 3, 5, and 9 months. They found that titers against tetanus toxoid and two pertussis antigens (toxin and fimbriae) were significantly lower after the accelerated schedule in infants in whom maternally derived antibody concentrations were higher. Therefore the new schedule "can have an inhibitory effect on the responses to immunisation against tetanus and pertussis."

The late fetus has some capacity to produce antibody on its own, as has been shown for meningococcal polysaccharide, streptococcal type B polysaccharide, and tetanus toxoid.

*Protective effects of breast-feeding.* Some important types of protection due to human breast milk are shown in Table C–3. Some specific infectious agents whose effects are reduced or prevented by breast milk include:

TABLE C–3.    Types of Protection Due to Breast-feeding

| |
|---|
| 1. Poorer conveyance of pathogens by human milk |
| 2. Contraceptive effect of breastfeeding |
| 3. Direct-acting host defense agents in human milk |
| 4. Host defense agents created in the recipient's alimentary tract by limited digestion of substrates in ingested human milk |
| 5. Promotors of protective microflora |
| 6. Antiinflammatory factors in human milk |
| 7. Immunostimulants in human milk |

*Source:* Goldman and Goldblum (1990).

1. Respiratory: respiratory syncytial virus; *Haemophilus influenzae; Streptococcus pneumoniae*
2. Enteric: Poliovirus, rotavirus, and other enteric viruses, *Escherichia coli; Shigella; Salmonella; Cryptosporidium* (Goldman and Goldblum 1990)

If a vaccine is proposed against any of the organisms listed, then breastfeeding must be taken into account in any field trial or proposed routine use. Saif (1985) showed in swine and cattle that maternal immunization against coronavirus and rotavirus led to the continual presence of milk–derived antibodies in the infant gut. Harmful infections were prevented but subclinical infections persisted, thereby soliciting an immune response from the young animal. Boosting the level of passive immunity may afford more complete protection from infection during the suckling period, but may result in an abrupt transition from "protected" to "susceptible" status during weaning.

Specific protective factors in breast milk, other than immunoglobulins, include white blood cells (macrophages, neutrophils, lymphocytes), lactoferrin, lactoperoxidase, lysozyme, antiinflammatory agents, and a variety of other molecules (Goldman and Goldblum 1990). The high levels of IgA in breast milk prevent infection in infants by blocking microbial attachment to the gut wall rather than by killing infectious agents directly. This protection is particularly important in the very young, since the intestinal wall is more permeable to antigens early in life. Breast milk may contain factors that stimulate the production of IgA by the infant (Björkstén 1986).

*Relative immunoresponsiveness in breast-fed and formula-fed infants.* Antibody response to vaccinations in healthy infants can vary with their usual diet. Zoppi et al. (1983) showed that responses to poliovirus, diphtheria, and tetanus toxoids were lower in breast-fed infants. In a similar study with hepatitis surface antigen (HBsAg) positive mothers, de Martino et al. (1987) administered hepatitis immune globulin to infants on the first day, and purified inactivated HBsAg (Hevac B, Institute Pasteur, Paris) during the second, fourth, and ninth month after birth. All infants seroconverted, but formula-fed infants showed significantly higher anti-HBs levels 3 months after the second injection. By 1 year of age the difference had become statistically insignificant. The temporary depression in anti-HBs antibodies in breast-fed infants may be a result of suppressive factors in milk, or immunologic tolerance in children who ingest the antigen.

*Stimulation of mammary glands by antigens present in the mother.* The lymphoid cells that make IgA in the Peyer's patches of the intestine can migrate to the glandular tissue of the breast, where they continue to make secretory IgA that is incorporated into the milk. Thus the infants ingest antibodies against pathogens that have already infected their mothers, and that may be likely to infect the children themselves. Parenteral cholera vaccination of mothers results in a boosting effect in cholera-specific IgA levels found in the milk. When lactating volunteers in Bangladesh were injected with anticholera toxoid and cholera whole-cell vaccine, IgA milk titers against anticholera toxoid rose in 5 of 6 volunteers, and 3 of those 5 showed increases in anticholera toxoid IgG titers. None of these results was seen in lactating women who received tetanus toxoid (Merson et al. (1980).

*Maternal-infant interaction in measles.* Measles represents a good case study of the interactions among the infective agent, host age, maternal antibody status, and vaccines (which are discussed in detail in chapter 5). If the mother has had or been immunized against measles, the very young infant is protected to a degree because of antibodies acquired from its mother. Lennon and Black (1986) reported that in both young and old mothers, maternal measles antibodies are concentrated by a factor of 1.7 across the placenta and have a mean half-life of 48 days. The child's initial, placentally derived titer is thus proportional to maternal titer. These passively acquired antibodies decline at a steady rate as the child matures and are further diluted because of increases in body size and blood volume, but as long as they are present the infant has some protection against measles. Harry and Ogunmekan (1981) in Nigeria noted that the percentage of infants with detectable measles antibody dropped from 100% at birth to 90% at 5 months, 50% at 7 months, and 42% at 8 months. In Congo, 95.8% of children of antibody-positive mothers showed detectable levels of serum antibodies at 8 to 11 weeks, but only 10.1% showed equivalent levels at 28 to 35 weeks of age (Dabis et al. 1989). In terms of the duration of antibody titer, a 2-fold difference in maternal antibody titer translates into an additional month and a half during which the infant will be protected from disease, but may also be resistant to vaccine.

Immunization with measles vaccine provokes lower titers than does the wild-type infection. As a result, children of vaccinated mothers tend to have lower initial titers and shorter windows of protection after birth, and will be responsive to live measles vaccine at an earlier age. Measles vaccines are usually considered to be effective at about 8 months of age, when the passive immunity transferred from the mother has declined sufficiently to avoid interference with the vaccine virus. However, a study in Haiti by Halsey et al. (1985) found better seroconversion rates among infants from mothers who had low titers; mothers with high measles titers were associated with infants who had difficulty converting.

## Age: The Effects of Aging

Aging depresses delayed-type hypersensitivity and proliferative response to both antigens and T-cell mitogens (Oyeyinka 1984). Serum IgE levels tend to decrease with age for women, but not for men. Females tend to show more autoantibodies with age, while males have higher levels of circulating immune complexes. For the elderly, the main determinant of depressed cellular immunity is age and age-related disease (decreased number of responding cells and a reduced ability of those cells to multiply). The elderly may also have reduced T-independent reactivity to bacterial polysaccharide antigens in a manner similar to very young children. In a malaria-endemic area of Mali, qualitative and quantitative differences in children and adults occur in the humoral response to certain malarial antigens (Fruh et al. 1991).

## Pregnancy

A mild cell-mediated immune deficiency occurs during pregnancy (Greenwood and Whittle 1981c). This temporary depression of immune response has been thought to be responsible for the prevalence of clinical malaria in pregnant women who were previously well.

## Ethnic and Genetic Factors in Immunity

In addition to nutrition and life experience, the genetic makeup of the local population can modify the efficacy of certain vaccines. It is not a simple matter to distinguish among the many potential influences on immunologic response. For example, the secretory IgA antibody response to cholera vaccine was compared between groups of ten lactating women from Sweden and Pakistan. Most Pakistani women were found to have significant preimmunization titers of specific IgA against cholera, whereas the Swedish women did not. A single subcutaneous injection of vaccine produced a significant boost in titer in 70% of milk and 45% of saliva samples from Pakistani women, but two injections 2 weeks apart failed to induce any response in milk or saliva in the Swedish women. This result was interpreted by Svennerholm et al. (1980) as arising from the priming effect of a previous exposure, but ethnic and genetic differences, if any, were not taken into account.

Two separate elements are involved in the genetics of immunologic responses in human populations:

1. Individuals may display inherent differences in their resistance or susceptibility to invasion, or establishment of an infection, by certain pathogens (modified after Wakelin and Blackwell 1988).

    A. Failure to gain access to the host's tissues because of impenetrable outer or inner surfaces. Included here are variations in structure, secretions, or characteristics of skin, and respiratory tract or intestinal mucosa.

    B. Absence of appropriate receptors for attachment and penetration of cell membranes. Each pathogen surface may possess a variety of sites that bind to particular determinants on host cells. The best known of these is the need of *Plasmodium vivax* to attach to the Duffy blood group antigen before it can penetrate the red cell surface. Most Africans lack this antigen and are therefore genetically insusceptible to infection with vivax maleria.

C. Failure to survive within the body because of naturally occurring lethal factors. Complement and related factors in blood may inhibit the development of certain protozoa.

D. Failure to undergo normal growth and development because of the absence of correct developmental stimuli or of inadequate nutritional supply. A good example is hemoglobin S in red blood cells, which is inhibitory to the development of malaria parasites.

2. Individuals differ in their inherent, genetically determined ability to respond immunologically to particular infections.

A. Cellular immunity. Destruction within tissues by granulocytes (e.g., eosinophils) and phagocytic cells such as macrophages.

B. Humoral immunity. The capability of lymphoid cells to respond to antigens is controlled by each individual's immune response (Ir) genes, which appear to be in the region of the major histocompatibility (MHC) gene complex on chromosome 6.

Immunocompetence, like all other characters, is largely inherited, and its relation to ethnic and racial makeup is still not well understood. Some associations between human leukocyte antigen (HLA) types and disease expression are unique to certain populations as a result of differing frequencies of HLA antigens. In other cases, the link between HLA and disease is consistent regardless of ethnic group. For example, HLA-B27 has a strong association with ankylosing spondylitis and Reiter's syndrome; B7 with multiple sclerosis; B8 with myasthenia gravis; DR2 is related to decreased incidence of juvenile diabetes mellitus, and so forth.

*Genetic differences in individual responsiveness.* One of the great questions of immunology is how individuals can express specific humoral immune responses to such a great variety of external antigens. Although the immunoglobulin (Ig) genes generate diversity by unique and complex mechanisms of selection and reassembly, the number of potential antibodies that can be made by any individual is not infinite.

Studies on inbred mice have revealed genetically determined differences, based on single substitutions in the MHC region, in ability to mount immune responses against complex antigen mixtures or single defined peptides. That humans function similarly is now beyond question, and for a vaccine field trial it is important to recognize that some individuals in some populations may be inherently inert to the vaccine. The ultimate fear is that these genetically determined differences in the host may mean that a vaccine that is protective in one individual may be suppressive or produce no response in another. The use of more sophisticated molecular genetic technology to analyze genetic variation in response to vaccine antigens is a priority area for future research.

Observations in humans in New Guinea and The Gambia (Carter et al. 1989; Good et al. 1988; Quakyi et al. 1989) have indeed revealed substantial proportions (to 40%) of individuals who do not appear to respond to specific sporozoite, merozoite, and gamete antigens from malarial parasites. These observations suggested that

> human genetic factors, not just the degree of exposure to parasite infection, are at least partly responsible for non-responsiveness in the adult population. . . . We interpret the widespread humoral non-responsiveness as evidence that many individuals do not recognize helper T cell epitopes from these critical malaria proteins. Our observations are relevant to vaccine development because these proteins may not be naturally immunogenic in many individuals, and vaccination with an entire protein will often result in no immune response (Quakyi et al. 1989).

The best evidence that control of immune response is associated with the major histocompatibility complex (MHC) of genes in humans comes from important studies on the inheritance of the ability to respond to hepatitis B surface antigen (HbsAg) vaccine. Nonresponders to this vaccine were associated with a certain histocompatibility leukocyte (HLA) type that lacks an immune response gene for HBsAg. In a familial study in which pairs of siblings had identical relevant HLA types, both members of eight pairs had similar antibody responses. Persons heterozygous for this particular HLA type always had higher antibody responses to HBsAg vaccine than their homozygous relatives. Response to HBsAg vaccine was shown to be inherited in a dominant fashion. The inability to respond to antigens is specific, because more than 95% of those who failed to the hepatitis vaccine did respond to tetanus toxoid (Kruskall et al. 1992).

## Sex

In West Africa the case-fatality rate for measles is significantly higher in girls than in boys. Measles infection contracted from a person of the opposite sex is more severe. Secondary cases infected by a child of the opposite sex had 2.4 times the risk of death than secondary cases infected by a child of the same sex (Aaby 1992).

Extreme sex differences in the occurrence of autoimmune diseases such as lupus erythematosis are well established. It seems reasonable that other immune responses may also differ between males and females. Infection with rubella virus (RV) results in different amounts and rates of production of various immunoglobulin subgroups in human males and females. "These differences suggest that there are hormonal and genetic influences on immune recognition of RV proteins that may be related to the increased incidence of rubella-associated arthropathy in females" (Mitchell et al. 1992).

## Other Factors

Several immunological "abnormalities" occur in healthy individuals in tropical countries, relative to those in industrialized countries (Greenwood and Whittle 1981a). Differences include higher serum levels of IgG, IgM, and IgE, prevalence of heterophile and Wasserman antibodies, and higher levels of immune complexes. Cellularly, lymphocyte subpopulations are altered.

Immunosuppressive drugs and pathological states naturally depress immune response and may activate latent infections.

Many authors have reported that oral polio vaccine often fails to induce strong seroconversion in infants in developing countries (Patriarca et al. 1991). Factors implicated include breast-feeding, enteroviruses, and gastric or salivary inhibitors. Data on breast-feeding are inconclusive. Swartz et al. (1972) studied more than 200 Israeli infants approximately 2 months old at the time of oral polio vaccine administration. Factors that influenced seroconversion were identified as: (1) season of administration—children vaccinated in the winter seroconverted more successfully; (2) socioeconomic class; and (3) presence of enteroviruses. The difference between winter and summer vaccination success was greater for children of low socioeconomic class and for those infected with other viruses. Therefore the optimal strategy may be a 3-dose administration during the cool season, given early in life before the invasion of enteroviruses, with an emphasis on low socioeconomic groups.

### Antibody-Dependent Enhancement (ADE) of Viral Diseases

*Dengue hemorrhagic fever.* Although most infections with dengue viruses are mild or even asymptomatic, dengue hemorrhagic fever and dengue shock syndrome (DHF/DSS) are severe illnesses that cause significant mortality in young children in Southeast Asia. Halstead (1970) suggested that the degree of infection appears to be regulated by the presence in the host of preexisting infection-enhancing antibodies (IEA), particularly those to a different serotype of dengue virus. In this view, the mechanism of ADE is that viruses, complexed with nonneutralizing antibodies, infect mononuclear phagocytes, resulting in a greater infection than that produced by the virus alone. Support has come from various experimental and epidemiological studies (e.g., Kliks et al. 1988), but the concept has been challenged (e.g., Rosen 1977, 1982).

The degree of ADE depends on the concentration of antibody. In initial infections, where there is no prior antibody, disease tends to be minimal to moderate; conversely, sufficiently high titers provide protection against viral proliferation and severe disease. However, at low to intermediate levels, antibody-dependent enhancement is observed. People can acquire dengue antibodies in two ways: through infection, or as transferred maternal antibodies:

> In contrast to childhood DHF/DSS, DHF/DSS among infants (< 12 months) is associated almost exclusively with the initial (primary) dengue infection. Since serologic surveys in Bangkok, Thailand, demonstrated that 90% to 100% of women at childbearing age circulated dengue antibodies, it has been surmised that the risk factor for DHF/DSS in this age group is passively acquired maternal antibody. Since women of childbearing age in Bangkok are usually immune to more than one dengue serotype, it was postulated that concentration of maternal antibodies

received at birth would protect infants from dengue infection. . . . The strong correlation observed between the mother's neutralizing antibody titers to DEN-2 virus and the age of their infant at onset of DHF/DSS is compatible with the expectation that maternal antibody protected the infant from DEN-2 infection for a period which was dependent upon the titer of DEN-2 neutralizing antibodies acquired at birth. . . . In the early stage of an infant's life (generally the first 6 months), maternal dengue neutralizing antibodies protect the infant from dengue infection. As the maternal IgG degrades, dengue neutralizing activity decreases below the protective level (generally at 7–8 months . . .). If an infant is exposed to DEN-2 virus during this high risk period DHF/DSS may develop; and beyond the critical 2-months period, further IgG degradation results in a decrease of IEA, now insufficient to mediate enhanced infection. When an infant of this age is exposed to DEN-2 virus, an asymptomatic infection occurs (Kliks et al. 1988).

Since enhancement of viremia by low antibody titer has been proposed as the mechanism for high rates of DHF/DSS in Southeast Asian infants, it is at least conceivable that a live vaccine could induce a similar effect under at least two sets of conditions. First, if a highly antigenic vaccine is injected into individuals with very low antibody levels, induction of ADE might be observed. Alternatively, a vaccine of low antigenicity may provoke antibody production too low for protection, but sufficient for induction of ADE in the event of later exposure to wild-type virus.

*Other flaviviruses.* Immune enhancement has been reported in several members of this group, including yellow fever, West Nile, rabies, and respiratory syncytial viruses. The possibility of reciprocal cross-enhancement (antibodies to one virus; infection with another) cannot be dismissed a priori.

*Poliovirus.* Children with immune systems suppressed by malaria, malnutrition, and measles may be particularly susceptible to paralytic poliomyelitis as a result of infection with wild poliovirus of otherwise low virulence. If this is so, then such children would also be considered more prone to paralysis following administration of oral poliovirus vaccine, even without reversion of vaccine virus to virulence. The use of baby foods in place of breast milk may decrease the protection against polio and other diseases previously provided by passive immunity through maternal antibodies.

Provocation of paralytic polio by injections has been mentioned previously. The disease may be associated with localized inflammation that allows poliovirus to enter nerve endings and affect the central nervous system. Such inflammation may be associated not only with a vaccine or antibiotic, but with unsterile needles and syringes often used in poor countries. If the poliovirus is of low virulence and antibody production is just adequate to prevent its spread, paralysis may be confined to the site of injection (Sutter et al. 1992), an observation presumed to explain the high rate of single-limb paralysis in the tropics.

## The Effect of Infection-Controlling Drugs on the Immune Response to Vaccines

Chloroquine is a particularly important drug in tropical developing countries because of its suppressive and curative properties against sensitive strains of malaria. However,

> chloroquine is known to disrupt lysosomal enzyme functions, impairing antigenic recognition and subsequent antibody production. It may also inhibit the generation of immunoglobulin-secreting lymphocytes by suppressing the secretion of interleukin-1 by monocytes and have other immunosuppressive effects (Grabenstein 1990).

Chloroquine inhibits the antibody response to killed rabies vaccine and apparently has the potential for inhibiting the replication of live vaccine viruses and subsequent immune response, apparently through disruption of macrophage presentation abilities (Pappaioanou et al. 1986). Tsai et al. (1986) conducted a brief study in which 17d yellow fever vaccine was administered to individuals taking 500 mg per week of prophylactic chloroquine, and to individuals who did not begin prophylaxis until 9 weeks after immunization. Although the difference in mean antibody titers was not significant, the authors theorize that chloroquine could adversely affect response to yellow fever vaccine through inhibition of antigen presentation by macrophages, or by direct inhibition of viral replication. Whole blood levels of the drug may be 10 times greater than plasma levels, but since these concentrations are centered in platelets and erythrocytes, they are unlikely to affect vaccine virus proliferation.

### Effect of Immunization on Metabolism and Nutritional Status

In a study carried out in nine villages in the Punjab, vaccinated children were compared with unvaccinated controls, excluding those with preexisting infections (Kielmann 1977). Children under 6 months vaccinated with BCG, smallpox, polio, and polio-DPT combined vaccines showed a significantly lower rate of weight gain, or even a drop in weight, compared to age-matched controls. These results imply that a weight drop after immunization may result from the infection induced by live agents. While a live antigen would be expected to bring about a more severe infection in a malnourished child than in a healthy one, it may be that a reduction in the nutritional health of an already compromised individual would render that child more susceptible to accompanying infections and less able to gain weight. If correct, these observations suggest that immunization of very young, malnourished children with live vaccines should be carried out with caution and with close monitoring of subsequent health status.

### Genetic Variability of the Target Organism

Many microorganisms, such as the viruses of measles and yellow fever, or the bacteria causing pertussis, are essentially the same everywhere in the world. Others, such as polio virus, occur in several widely distributed stable serotypes. Still others, such as human immunodeficiency virus, respiratory syncytial virus, and influenza, are genetically so variable that it is difficult to make appropriate vaccines for general use.

*Special problems of antiparasite vaccines.* Parasitic organisms present far greater complexities. First, most parasitic protozoa and helminths (worms) occur naturally in numerous geographic variants adapted to local climate, vectors, intermediate hosts, and other conditions. Second, they generally go through complex life cycles that demonstrate substantial morphological, biochemical, and behavioral differences from one stage to the next. Immunologic or pharmacologic tools may be useful against one form, but inactive against others. For example, the drug Ivermectin kills microfilariae (the transmission form), but not adults, of *Onchocerca volvulus*, the worm causing river blindness in West Africa and parts of Central and South America. Third, the number of genes present is far greater in helminths than in viruses or bacteria. Schistosome parasites, for instance, possess about 10% as much nuclear DNA as do mammals. Not only is the biochemical makeup of these organisms correspondingly more complex and variable, but the likelihood that protective immunity will develop against any particular gene product is relatively less than in microbes. Accordingly, parasitic organisms possess great genetic adaptability and generate a broad array of pathways by which to escape from biochemical or immunologic threats.

*A case study: Malaria.* Among protozoa, the surface glycoproteins of African trypanosomes, the cause of sleeping sickness, are notoriously changeable. Similarly, the capacity of malarial parasites to become resistant to drugs has given rise to numerous distinguishable phenotypes in various parts of the world. The mechanisms to generate and maintain the underlying genetic polymorphisms are inherent in these organisms. It is possible that a vaccine, in the same way as antimalarial drugs, may loss efficacy over time because selection pressure leads to evolution of resistant populations.

A great deal of effort has been expended in the attempt to produce a vaccine against *Plasmodium falciparum*, the cause of malignant tertian malaria. Two basic elements can determine a generally useful vaccine: first, the antigenic characteristics of the malaria parasite should remain stable over time and in a wide geographic area; second, the ability to mount a protective immune response should be universal, or nearly so, among human beings. The antigenic structure of *P. falciparum* has been intensively studied, and various stages of the parasite have been considered as sources of the subunits from which to make the vaccine. A great deal of early work was done on the sporozoite, which is transmitted from the mosquito to the human host and is exposed in the bloodstream for a brief period. Although some natural variation occurs within the sporozoite surface from one isolate to another, the significance of this observation is still not completely understood. The merozoite stage, which emerges from the ruptured red blood cell, is similarly accessible to immune attack, and its antigens would also seem to be prime candidates for a vaccine. Smythe et al. (1991) studied the structure of a particular surface antigen (MSA-2) from merozoites of 44 different geographic strains of *P. falciparum,* finding considerable diversity among different isolates, although there were also common features in all. Certa (1991) discussed the genetic diversity of RESA (ring-infected erythrocyte surface antigen), found to be protective against *P. falciparum* in preliminary trials in monkeys:

> The best protection, however, was induced by a highly repetitive sequence which is subject to antigenic variation and therefore of limited use. More important, cultured parasites were discovered that had abolished expression of the RESA gene as a result of a chromosomal deletion. This may occur in nature and it is therefore evident that vaccine candidates against malaria cannot be selected on the basis of a protective immune reponse since their efficiency may only last for a short period owing to the genetic flexibility of *P. falciparum.*

In Papua New Guinea, studies preliminary to proposed malaria vaccine field trials employed strain-specific sequences of a particular surface antigen from a particular serotype of *Plasmodium falciparum,* which has been expressed in *Escherichia coli.* In villages as little as 10 kilometers apart, in which about 30% of the residents had *P. falciparum* malaria, the frequency of occurrence of that particular serotype varied from 0% to 42% at any one time, and also varied considerably within the same village from year to year. On the other hand, some features of the malarial surface are extremely stable, and these could be exploited for vaccines, provided that a good protective response is obtained (Forsyth et al. 1988).

# References

Aaby P. 1992. Influence of cross-sex transmission on measles mortality in rural Senegal. *Lancet* 340:388–391.

Aaby P, Bukh J, Leerhoy J, et al. 1986. Vaccinated children get milder measles infection: a community study from Guinea-Bissau. *Journal of Infectious Diseases* 154:858–863.

Aaby P, Jensen TG, Hansen HL, et al. 1988. Trial of High-dose Edmonston-Zagreb measles vaccine in Guinea-Bissau: protective efficacy. *Lancet* 2(8615):809–811.

Abdel-Wahab MF, Esmat G, Narooz SI, et al. 1990. Sonographic studies of schoolchildren in a village endemic for Schistosoma mansoni. *Transactions of the Royal Society of Tropical Medicine and Hygiene* 84:69–73.

Ada GL. 1986. Antigen presentation and enhancement of immunity: an introduction. In F Brown, RM Channock, RA Lerner, eds. *Vaccines 86*. Cold Spring Harbor. CSH Laboratory. 105–108.

Adityanjee. 1986. Informed consent: issues involved for developing countries. *Medical Science and the Law* 26:305–307.

Advisory Committee for Health Research. 1986. Enhancement of transfer of technology to developing countries with special reference to health. Geneva. WHO Document ACHR28/86.5. 32 pp.

Agarwal A. 1978. *Drugs and the Third World*. London: Earthscan Publications.

Agócs MM, Markowitz LE, Straub I, Dömök I. 1992. The 1988–1989 measles epidemic in Hungary: assessment of vaccine failure. *International Journal of Epidemiology* 21:1007–1013.

Alexander M. 1985. Spread of organisms with novel genotypes. In AH Teich, MA Levin, JH Pace, eds. *Biotechnology and the Environment*. Washington D.C.: American Association for the Advancement of Science. 115–127.

Almond JW, Cann AJ. 1984. Attenuation. In IM Roitt, ed. *Immune Intervention Vol 1. New Trends in Vaccines*. London: Academic Press. 13–56.

Alubo SO. 1990. Debt crisis, health and health services in Africa. *Social Science and Medicine* 31:639–647.

Amato I. 1993. Cancer vaccine slated for tests. *Science* 259:758.

Anderson RM. 1988. Epidemiological models and predictions. *Tropical and Geographical Medicine* 40:530–539.

Anderson RM, May RM. 1982. The logic of vaccination. *New Scientist* 96:410–415.

———. 1983. Vaccination against rubella and measles: quantitative investigations of different policies. *Journal of Hygiene* 90:259–325.

Anonymous. 1930. Verschiedenes. *Zeitschrift für Tuberkulose* 57:125–126.

———. 1991a. Essential Drugs and Vaccines. *World Health Statistics Annual for 1990*. Geneva. World Health Organization. 19–21.

———. 1991b. Outbreaks of rubella among the Amish—United States, 1991. *MMWR. Morbidity and Mortality Weekly Report* 40:264–65.

Assaad F. 1979. Poliomyelitis vaccination. Benefit versus risk. *Developments in Biological Standardization* 43:141–150.

Ashford RW, Desjeux P, deRaadt P. 1992. Estimation of population at risk of infection and number of cases of leishmaniasis. *Parasitology Today* 8:104–105.

Axnick NW, Shavell SM, Witte JJ. 1969. Benefits due to immunization against measles. *Public Health Reports* 84:673–680.

Bagchi K, Mohanram M, Vinodini R. 1980. Humoral immune response in children with iron-deficiency anaemia. *British Medical Journal* 280:1249–1250.

Bancroft WH. 1992. Hepatitis A vaccine. *New England Journal of Medicine* 327:488–490.

Banerji D. 1990. Prescription without diagnosis. Report of Commission on Health Research for Development. *Economic and Political Weekly* (Bombay). Dec 26:2823–2825.

Bankowski Z, Bryant JH, Last, JM, eds. 1991. *Ethics and Epidemiology: International Guidelines*. Geneva, Switzerland. Council for International Organizations of Medical Sciences (CIOMS). 163 + 28 pp.

Banta D, Andreasen PB. 1990. The political dimension in health care technology assessment programs. *International Journal of Technology Assessment in Health Care* 6:115–123.

Barnum HD, Tarantola D, Setiady T. 1980. Cost-effectiveness of an immunization program in Indonesia. *Bulletin of the World Health Organization* 58:499–503.

Barron PM, Buch E, Behr G, Crisp NG. 1987. Mass immunization campaigns—do they solve the problem? *South African Medical Journal* 72:321–322.

Barry A. 1989. Rural resistance to technological innovation. Causes and solutions. *Industry Africa* 1:3–5.

Bart KJ, Lin KF. 1990. Vaccine-preventable disease and immunization in the developing world. *Pediatric Clinics of North America* 37:735–756.

Barton JH. 1989. Legal trends and agricultural biotechnology: effects on developing countries. *Trends in Biotechnology* 7:264–268.

Basch PF. 1990a. Biomedical innovation and world health. *Perspectives in Biology and Medicine* 33:501–508.

———. 1990b. *Textbook of International Health*. New York: Oxford University Press.

———. 1993a. Technology transfer and the delivery of health care. In JR Boldú, JR de la Fuente, eds. *Science Policy in Developing Countries: The Case of Mexico*. Mexico City: Fondo de Cultura Económica. 79–91.

———. 1993b. Antischistosomal vaccines: beyond the laboratory. *Transactions of the Royal Society of Tropical Medicine and Hygiene*. 87:589–92.

Baxby D, Paoletti E. 1992. Potential use of non-replicating vectors as recombinant vaccines. *Vaccine* 10:8–9.

Baumgaertner JC, Hogan R, Born C, et al. 1985. Contact spread of vaccinia from a national guard vaccinee—Wisconsin. *Journal of the American Medical Association* 253:2348.

Beale AJ. 1990. Polio vaccines: time for a change in immunization policy? *Lancet* 335:839–842.

Bektimirov T, Lambert PH, Torrigiani G. 1990. Vaccine development perspectives of the World Health Organization. *Journal of Medical Virology* 31:62–64.

Belcher DW, Nicholas DD, Ofosu-Amaah S, et al. 1978. A mass immunization campaign in rural Ghana: factors affecting participation. *Public Health Reports* 93:170–176.

Bellanti JA, Fishman HJ, Wientzen RL. 1987. Adverse reactions to vaccines. *Immunology and Allergy Clinics of North America* 7:423–445.

Bernard KW, Fishbein DB. 1991. Pre-exposure rabies prophylaxis for travellers: are the benefits worth the cost? *Vaccine* 9:833–836.

Bertrand M. 1989. *The Third Generation World Organization*. Dordrecht, Netherlands: Martinus Nijhoff Publishers. 217 pp.

Bhaskaram P, Madhusudan J, Radhakrishna V, Raj S. 1986. Immunological response to measles vaccination in poor communities. *Human Nutrition: Clinical Nutrition* 40C:295–299.

Bialy H. 1988. New Third-World production initiative. *Bio/Technology* 6:255.

Bidwai P. 1987. Biotechnology: the second colonialization of the Third World? *Times of India* 19 April.

Bjerregaard P. 1991. Economic analysis of immunization programmes. *Scandinavian Journal of Social Medicine* 46 Supplement:1115–1119.

Björkstén B. 1986. Immune responses to ingested antigens in relation to feeding pattern in childhood. *Annals of Allergy* 57:143–146.

Blackwelder WC, Storsaeter J, Olin P, Hallander HO. 1991. Acellular pertussis vaccines. Efficacy and evaluation of clinical case definitions. *American Journal of Diseases of Children* 145:1285–1289.

Bloom BR. 1986. Learning from leprosy: a perspective on immunology and the Third World. *Journal of Immunology* 137:i–x.

———. 1989. Vaccines for the Third World. *Nature* 342:115–120.

Bloom BR, Mehra V. 1984. Vaccine strategy for the eradication of leprosy. In R. Bell, G Torrigiani, eds. *New Approaches to Vaccine Development*. Proceedings of a Meeting Organized by the World Health Organization in Geneva. 17–20 October 1983. Basel: Schwabe and Co. AG. 368–389.

Bomford R. 1984. Relative adjuvant efficacy of Al(OH)$_3$ and saponin is related to the immunogenicity of the antigen. *International Archives of Allergy and Applied Immunology* 75:280–281.

———. 1989. Adjuvants for anti-parasite vaccines. *Parasitology Today* 5:41–46.

Booy R, Aitken SJM, Taylor S, et al. 1992. Immunogenisity of combined diphtheria, tetanus and pertussis vaccine given at 2, 3, and 4 months versus 3, 5, and 9 months of age. *Lancet* 339:507–510.

Borgoño JM, Corey G. 1978. The Chilean experience with antipoliomyelitis vaccination. *Developments in Biological Standardization* 41:141–148.

Brenzel L. 1989. *The Cost of EPI. A Review of Cost and Cost-Effectiveness Studies 1979–1987*. Arlington, Va.: John Snow, Inc. REACH Project. 101 pp.

———. 1990. *The Costs of EPI: Lessons Learned from Cost and Cost-Effectiveness Studies of Immunization Programs*. Arlington, Va.: John Snow, Inc. REACH Project. 89 pp.

Breo DL. 1993. The U.S. race to 'cure' AIDS—at '4' on a scale of 10, says Dr. Fauci. *JAMA* 269:2898.

Brés P. 1979. Benefit versus risk factors in immunization against yellow fever. *Developments in Biological Standardization* 43:297–304.

Bright JR. 1978. *Practical Technology Forecasting*. Austin, Tex.: Industrial Management Center. 351 pp.

Brorsson B, Wall S. 1985. *Assessment of Medical Technology—Problems and Methods*. Stockholm: Swedish Medical Research Council. 128 pp.

Brown KR, Douglas RG. 1992. New challenges in quality control and licensure: regulation. Presented at Conference on Vaccines and Public Health: Assessing Technologies and Global Policies for the Children's Vaccine Initiative, Bethesda, Maryland, Nov. 1992.

Bryan JP, Sjogran MH, Perine PL, Legters LJ. 1992. Low-dose intradermal and intramuscular vaccination against hepatitis B. *Clinical Infectious Diseases* 14:697–707.

Burrill GS, Roberts WJ. 1992. Biotechnology and economic development: the winning formula. *Bio/Technology* 10:647–653.

Caldwell LK. 1988. International aspects of biotechnology. *MIRCEN Journal* 4:245–248.

Canadian Task Force on the Periodic Health Examination. 1979. The periodic health examination. *Canadian Medical Association Journal* 121:1193–1254.

Carlini EA, Herxheimer A. 1991. Brazil: Pharmaceutical industry fears informed criticism. *Lancet* 337:724.

Carter R, Graves PM, Quakyi IA, Good MF. 1989. Restricted or absent immune responses in human populations to *Plasmodium falciparum* gamete antigens that are targets of malaria transmission-blocking antibodies. *Journal of Experimental Medicine* 169:135–147.

Certa V. 1991. Malaria Vaccine. *Experientia* 47:157–163.

Chandra RK. 1975. Reduced secretory antibody response to live attenuated measles and poliovirus vaccines in malnourished children. *British Medical Journal* 2, 583–585.

———. 1988. Nutrition and immunity. *Tropical and Geographical Medicine* 40:s46–s51.

———. 1991. Immunocompetence is a sensitive and functional barometer of nutritional status. *Acta Paediatrica Scandinavica* Suppl. 347:129–132.

Chetley A. 1990. *A Healthy Business? World Health and the Pharmaceutical Industry*. London and Atlantic Highlands, N.J.: Zed Books.

Children's Vaccine Initiative. 1991. The views of vaccine producers (comments). Report of the First Meeting of the Consultative Group. Geneva; 16–18 December 1991. Document CVI/91.5. pp 10–11.

Christakis NA. 1988. The ethical design of an AIDS vaccine trial in Africa. *Hastings Center Report* 18:31–37.

Clemens JD, Shapiro ED. 1984. Resolving the pneumococcal vaccine controversy: are there alternatives to randomized clinical trials. *Reviews of Infectious Diseases* 6:589–600.

Coates JF. 1976. Technology assessment—a tool kit. *Chemtech* (June) 372–383.

Cocchetto D, Nardi RV. 1986. Benefit-risk assessment of investigational drugs: current methodology, limitations, and alternative approaches. *Pharmacotherapy* 6:286–303.

Cohen H, Hofman B, Brouwer R, et al. 1973. Combined inactivated vaccines. *Symposia Series in Immunobiological Standardization* 22:133–141.

Commission on Health Research for Development. 1990. *Health Research. Essential Link to Equity in Development*. New York: Oxford University Press. 136 p.

Comstock GW. 1990. Vaccine evaluation by case-control or prospective studies. *American Journal of Epidemiology* 131:205–207.

Contractor FJ, Sagafi-Nejad T. 1981. International technology transfer: major issues and policy responses. *Journal of International Business Studies* Fall:113–135.

Cook AG. 1989. Patents as non-tariff trade barriers. *Trends in Biotechnology* 7:258–263.

Cornia Ga, Jolly R, Stewart F. 1987. *Adjustment With a Human Face*. Oxford, UK: Clarendon Press. 2 vols.

Council for International Organizations of Medical Sciences and WHO. 1982. Proposed international guidelines for biomedical research involving human subjects. Geneva.

Creese AL. 1986. Cost effectiveness of potential immunization interventions against diarrheal disease. *Social Science and Medicine* 23:231–240.

Crispen R. 1989. History of BCG and its substrains. *Progress in Clinical and Biological Research* 310:35–50.

Crouch EAC, Wilson R. 1982. *Risk/Benefit Analysis*. Cambridge, Mass.: Ballinger Publishing Co.

Curtis T. 1992. The origin of AIDS. *Rolling Stone*. March 19:54, 56–57, 59–60, 106, 108.

Dabis F, Waldman RJ, Mann GF, et al. 1989. Loss of maternal measles antibody during infancy in an African city. *International Journal of Epidemiology* 18:264–268.

da Silva EJ. 1983. Biotechnology in development of cooperation: a developing country's view. In PA van Hemert, HJM Lelieveld, JWM la Rivere, eds. *Biotechnology in Developing Countries*. Delft, Netherlands: Delft University Press. 19–57.

da Silva EJ, Sasson A. 1989. Promises of biotechnologies and the developing countries. *MIRCEN Journal* 5:115–118.

DaVilla G, Piazza M, Iorio R, et al. 1992. A pilot model of vaccination against hepatitis B virus suitable for mass vaccination campaigns in hyperendemic areas. *Journal of Medical Virology* 36:274–278.

Daynes RA. 1992. Prevention and reversal of age-associated changes in immunologic responses by supplemental dehydroepiandrosterone sulfate therapy. *Aging Immunology and Infectious Disease* 3:135–154.

Delafuente JC. 1991. Nutrients and immune responses. *Rheumatic Disease Clinics of North America* 17:203–212.

de Martino M, Resti M, Appendino C. Vierucci A. 1987. Different degree of antibody response to hepatitis B virus vaccine in breast- and formula-feed infants born to HBsAg-positive mothers. *Journal of Pediatric Gastroenterology and Nutrition* 6:208–211.

DeQuadros C, Carrasco P, Olive J-M. Desired field performance characteristics of new improved vaccines for the developing world. Presented at Conference on Vaccines and Public Health: Assessing Technologies and Global Policies for the Children's Vaccine Initiative, Bethesda, Maryland, Nov. 1992.

Desowitz RS. 1991. *The Malaria Capers*. New York: W.W. Norton and Company.

De Wilde M. 1987. Vaccine development within industry. *Acta Tropica 44 Suppl* 12:104–107.

Dingell JD. 1985. Benefits for the developing world. *Bio/Technology*. 3:752.

Domínguez-Ugá MA. 1988. Economic analysis of the vaccination strategies adopted in Brazil in 1982. *Bulletin of the Pan American Health Organization* 22:250–268.

Dunlop JM. 1988. The history of immunisation. *Public Health* 102:199–208.

Dupuy JM, Freidel L. 1990. Lag between discovery and production of new vaccines for the developing world. *Lancet* 336:733–734.

Dyke T. 1991. Public health aspects of a refugee camp. *Journal of the Royal Society of Health* 111:101–104.

Edgington SM. 1992. Biotech vaccines: problematic promise. *Bio/Technology* 10:763–766.

Ekunwe EO. 1984. Expanding immunization coverage through improved clinic procedures. *World Health Forum* 5:361–363.

Eldridge JH, Staas JK, Meulbroek JA, et al. 1991. Biodegradable microspheres as a vaccine delivery system. *Molecular Immunology* 28:267–294.

Engelhardt HT Jr. 1988. Diagnosing well and treating prudently: Randomized clinical trials and the problem of knowing truly. In SF Spicker, I Alon, A de Vries, HT Engelhardt Jr, eds. *The Use of Human Beings in Research*. Dordrecht, Netherlands: Kleuver Academic Publishers. 123–141.

Eskola J, Käyhty H, Takala AK, et al. 1990. A randomized, prospective field trial of a conjugate vaccine in the protection of infants and young children against invasive *Haemophilus influenzae* type b disease. *New England Journal of Medicine* 323:1381–1387.

Esrey SA, Potash JB, Roberts L, Shiff C. 1991. Effects of improved water supply and sanitation on ascariasis, diarrhoea, dracunculiasis, hookworm infection, schistosomiasis, and trachoma. *Bulletin of the World Health Organization* 69:609–621.

Evans AS. 1985. The eradication of communicable diseases: myth or reality? *American Journal of Epidemiology* 122:199–207.

———. 1989. Surveillance and seroepidemiology. In AS Evans, ed. *Viral Infections of Humans* 3d ed. New York: Plenum Medical Book Co. 51–73.

Expanded Programme on Immunization. 1992a. *Report of the 14th Global Advisory Group 14–18 October 1991 Antalya, Turkey*. Geneva. WHO Document WHO/EPI/GEN/92.1. 72 pp.

———. 1992b. *EPI For the 1990s*. Geneva. WHO Document WHO/EPI/GEN/92.2. 15 pp.

Fagan EA, Williams R. 1987. Hepatitis vaccination (review). *British Journal of Clinical Practice* 41:569–576.

Felsenfeld O, Wolf RH, Gyr K, et al. 1973. Simultaneous vaccination against cholera and yellow fever. *Lancet* 1(7801):457–458.

Fendall R. 1985. Myths and misconceptions in primary health care: lessons from experience. *Third World Planning Review* 7:307–322.

———. 1987. The integration of vertical programmes into primary health care. *Third World Planning Review* 9:275–284.

Fine PEM. 1989. The BCG story: Lessons from the past and implications for the future. *Reviews of Infectious Diseases* 11 Suppl 2:S353–S359.

Fine PEM. Clarkson JA. 1986. Individual versus public priorities in the determination of optimal vaccination policies. *American Journal of Epidemiology* 124:1012–1020.

Fischoff B, Cox LA Jr. 1986. Conceptual framework for regulatory benefits assessment. In JD Bentkover, VT Covello, J Mumpower, eds. *Benefits Assessment: The State of the Art*. Dordrecht, Netherlands: D. Reidel Publ. 51–84.

Fischoff B, Slovic P, Lichtenstein S. 1980. Labile values: a challenge for risk assessment. In J Conrad, ed. *Society, Technology and Risk Assessment*. London: Academic Press. 57–66.

Forsyth KP, Anders RF, Kemp DJ, Alpers MP. 1988. New approaches to the serotypic analysis of the epidemiology of *Plasmodium falciparum*. *Philosophical Transactions of the Royal Society of London. Series B: Biological Sciences* 321:485–493.

Foster SO. 1987. Unpublished Memo: Field Report on Supervision of EPI Team. Atlanta, Ga.: Centers for Disease Control. 19pp.

Fox JL 1987a. Three recombinant vaccine tests stir debate. *Bio/Technology* 5:13–14.

———. 1987b. Public opinion: sense vs. sensibility. *Bio/Technology* 5:14.

Frank V. 1985. A general introduction to the new biotechnology business. In RA Bohrer, ed. *From Research to Revolution*. Littleton, Col.: Fred B. Rothman. 13–20.

Freeman P. Robbins A. 1991. The elusive promise of vaccines. *The American Prospect 1* (Winter):80–90.

Frenk J, Bobadilla JL, Sepúlveda J, López-Cervantes M. 1989. Health transition in middle-income countries: new challenges for health care. *Health Policy and Planning* 4:29–39.

Frerichs RR. 1991. Epidemiologic surveillance in developing countries. *Annual Review of Public Health* 12:257–280.

Fruh K, Doumbo O, Muller HM, et al. 1991. Human antibody response to the major merozoite surface antigen of *Plasmodium falciparum* is strain specific and short lived. *Infection and Immunity* 59:1319–1324.

Fuchs VR. 1974. *Who Shall Live?* New York: Basic Books. 168 pp.

Fuchs VR, Garber AM. 1990. The new technology assessment. *New England Journal of Medicine* 323:673–677.

Galazka A. 1989. Stability of Vaccines. Geneva: WHO Document WHO/EPI/GEN/89.8. 57 pp.

Galazka AM, Lauer BA, Henderson RH, Keja AJ. 1984. Indications and contraindications for vaccines used in the Expanded Programme on Immunization. *Bulletin of the World Health Organization* 62:357–366.

Garenne M, Leroy O, Beau J-P, Sene I. 1991. Child mortality after high-titre measles vaccines: prospective study in Senegal. *Lancet* 338:903–907.

Gershwin ME, Keen CL, Mareschi J-P, Fletcher MP. 1991. Trace metal nutrition and the immune response. *Comprehensive Therapy* 17:27–34.

Giammanco G, DeGrandi V, Lupo L, et al. 1988. Interference with oral poliovirus vaccine on RIT4237 oral rotavirus vaccine. *European Journal of Epidemiology* 4:121–123.

Gibbs, JN. 1987. Exporting biotechnology products: a look at the issues. *Bio/Technology* 5:46–51.

Gladwell, M. 1991. AIDS forecast focuses on Third World. *Washington Post,* 18 June.

Goldman AS, Goldblum RM. 1990. Human milk: immunologic–nutritional relationships. *Annals of the New York Academy of Sciences* 587:236–245.

Good MF, Miller LH, Kumar S, et al. 1988. Limited immunological recognition of critical malaria vaccine candidate antigens. *Science* 242:574–577.

Grabenstein JD. 1990. Drug interactions involving immunologic agents. Part I. Vaccine-vaccine, vaccine-immunoglobulin, and vaccine-drug interactions. *DICP* 24:67–81.

Grant JP. 1989. *The State of the World's Children 1989.* New York: Oxford University Press. Published for UNICEF.

Greenwood BM, Bradley-Moore AM, Bradley AK. 1986. The immune response to vaccination in undernourished and well-nourished Nigerian children. *Annals of Tropical Medicine and Parasitology* 80:537–544.

Greenwood BM. Greenwood AM, Bradley AK, et al. 1987. Deaths in infancy and early childhood in a well-vaccinated, rural West African population. *Annals of Tropical Paediatrics* 7:91–99.

Greenwood BM, Wali SS. 1980. Control of meningococcal infection in the African meningitis belt by selective vaccination. *Lancet* 1(8171):729–732.

Greenwood BM, Whittle HC. 1981a. Immunological changes in healthy individuals living in the tropics. In BM Greenwood, HC Whittle, eds. *Immunology of Medicine in the Tropics.* London: Edward Arnold. 1–20.

———. 1981b. The Immune Response to infection. In BM Greenwood, HC Whittle, eds. *Immunology of Medicine in the Tropics.* London: Edward Arnold. 21–61.

Greenwood BM, Whittle HC. 1981c. Nutrition, infection and Immunity. In BM Greenwood, HC Whittle, eds. *Immunology of Medicine in the Tropics.* London: Edward Arnold. 178–210.

———. 1981d. Immunization. In BM Greenwood, HC Whittle, eds. *Immunology of Medicine in the Tropics.* London: Edward Arnold. 246–283.

Guillaume JC, Saiag P, Wechsler J, et al. 1991. Vaccinia from recombinant virus expressing HIV genes. *Lancet* 337:1034–1035.

Guyer B, McBean AM. 1981. The epidemiology and control of measles in Yaoundé, Cameroun, 1968–1975. *International Journal of Epidemiology* 10:263–269.

Gwatkin DR. 1980. How many die? A set of demographic estimates of the annual number of infant and child deaths in the world. *American Journal of Public Health* 70:1286–1289.

Haaga JG. 1983. *Cost effectiveness and cost benefit analysis of immunization programs in developing countries: a review of the literature.* International Federation of Pharmaceutical Manufacturers Associations. Annex III, Document SV67. 58 pp.

Hafez M, Aref GH, Mehareb SW, et al. 1977. Antibody production and complement system in protein energy malnutrition. *Journal of Tropical Medicine and Hygiene.* 80:36–39.

Hall AJ, Aaby P. 1990. Tropical trials and tribulations. *International Journal of Epidemiology* 19:777–781.

Hall AJ, Greenwood BM, Whittle H. 1990. Modern Vaccines. Practice in developing countries. *Lancet* 335:774–777.

Halsey NA, Boulos R, Mode F, et al. 1985. Response to measles vaccine in Haitian infants 6 to 12 months old. Influence of maternal antibodies. *New England Journal of Medicine* 313:544–549.

Halstead SB. 1970. Observations related to pathogenesis of dengue hemorrhagic fever. VI. Hypotheses and discussion. *Yale Journal of Biology and Medicine* 42:350–362.

Halstead SB, Tugwell P, Bennett K. 1991. The International Clinical Epidemiology Network (INCLEN): a progress report. *Journal of Clinical Epidemiology* 44:579–589.

Hancock G. 1989. *Lords of Poverty. The Power, Prestige, and Corruption of the International Aid Business.* New York: The Atlantic Monthly Press. 234 pp.

Harris TJR. 1984. Gene Cloning in vaccine research. In IM Roitt, ed. *Immune Intervention. Vol. 1. New Trends in Vaccines.* London: Academic Press. 57–92.

Harry TO, Ogunmekan DA. 1981. Optimal age for vaccinating Nigerian children against measles. I. Neonatal antibody profile and subsequent susceptibility to measles. *Tropical and Geographical Medicine* 33:375–378.

Harte PG, Cooke A, Playfair JH. 1983. Specific monoclonal IgM is a potent adjuvant in murine vaccination. *Nature* 302:256–258.

Harvey PH, Keymer AG, May RM. 1988. Evolving control of diseases. *Nature* 332:680–681.

Hathout H, Al-Nakib W, Lilley H, et al. 1978. Seroepidemiology of rubella in Kuwait: an alternative vaccination policy. *International Journal of Epidemiology* 7:49–53.

Heath AM, Playfair JHL. 1992. Cytokines as immunological adjuvants. *Vaccine* 10:427–434.

Heggenhougen HK, Clements J. 1987. *Acceptability of Childhood Immunization. Social Science Perspectives: A Review and Annotated Bibliography.* London: London School of Hygiene and Tropical Medicine. Evaluation and Planning Centre for Health Care. 97 pp.

———. 1990. An anthropological perspective on the acceptability of immunization services. *Scandinavian Journal of Infectious Diseases Supplement* 76:20–31.

Henderson R. 1984a. The expanded programme on immunization of the World Health Organization. Reviews of Infectious Diseases 6:S475–S479.

――――. 1984b. An example of vaccine application: the Expanded Programme on Immunization. In R Bell, G Torrigiani, eds. *New Approaches to Vaccine Development.* Proceedings of a meeting organized by the World Health Organization in Geneva. 17–20 October 1983. Basel: Schwabe and Co. AG. 506–513.

――――. 1990. Vaccinations in the health strategies of the developing countries. *Scandinavian Journal of Infectious Diseases* 76 Supplement: 7–14.

Henderson RH, Sundaresan T. 1982. Cluster sampling to assess immunization coverage: A review of experience with a simplified sampling method. *Bulletin of the World Health Organization* 60:253–260.

Hilleman MR. 1986a. The science of vaccines in present and future perspective. *Medical Journal of Australia* 144:360–364.

――――. 1986b. Vaccinology in practical perspective. *Developments in Biological Standardization* 63:5–13.

――――. 1989. Improving the heat stability of vaccines: problems, needs, and approaches. *Reviews of Infectious Diseases* 11 Suppl 3:S613–S616.

Hilleman MR, Buynak EB, Weibel RE, Villarejos VM. 1973. Immune responses and duration of immunity following combined live virus vaccines. *Symposia Series in Immunobiological Standardization* 22:145–158.

Hofer MA. 1991. The protection of intellectual property in biotechnology. In RD Ono, ed. The *Business of Biotechnology.* Stoneham, Mass.: Butterworth-Heineman. 151–167.

Homeida M, Ahmed S, Dafalla A, et al. 1988. Morbidity associated with *Schistosoma mansoni* infection as determined by ultrasound: a study in Gezira, Sudan. *American Journal of Tropical Medicine and Hygiene* 39:196–201.

Homma A, Knouss R. 1992. Transfer of vaccine technology to developing countries. The Latin American experience. Contribution No. 13 from: *Vaccines and Public Health: Assessing Technologies and Global Policies for the CVI.* Conference at the National Institutes of Health, Bethesda, Maryland, November 5 and 6, 1992. 29 pages.

Houston S, Fanning A, Soskolne CL, Fraser N. 1990. The effectiveness of *Bacillus Calmette-Guérin* (BCG) vaccination against tuberculosis. A case-control study in treaty Indians, Alberta, Canada. *American Journal of Epidemiology* 131:340–348.

Hovi T. Remaining problems before eradication of poliomyelitis can be accomplished. *Progress in Medical Virology* 38:69–95.

Howson CP, Howe CJ, Fineberg HV, eds. 1991. *Adverse Effects of Pertussis and Rubella Vaccines. A Report of the Committee to Review the Adverse Consequences of Pertussis and Rubella vaccines.* Washington D.C.: National Academy Press.

Igelhart JK. 1987. Compensating children with vaccine-related injuries. *New England Journal of Medicine* 316:1283–1288.

Imam IZ. 1985. Immunization and primary health care in Egypt. *Tropical and Geographical Medicine* 37:565–566.

Immunization Practices Advisory Committee. 1989. Recommendations of the Immunization Practices Advisory Committee (ACIP): general recommendations on immunization. *MMWR Morbidity and Mortality Weekly Report* 38:205–214, 219–228.

Immunization Update. 1992. Public Vaccine Costs to Immunize a Child. United States. 1978–1992. Berkeley. State of California Department of Health Services Immunization Unit. Personal Communication.

Imperato PJ. 1975. *A Wind in Africa.* St. Louis, Mo.: Warren H Green, Inc. 363 pp.

――――. 1985. The potential of diagnostics for improving community health in less developed countries. *Journal of Community Health* 10:201–206.

Insel RA. 1988. Maternal immunization to prevent neonatal infections. *New England Journal of Medicine* 319:1219–1220.

Institute of Medicine. 1986. New Vaccine Development. Establishing Priorities. *Volume I. Diseases of Importance in the United States. Volume II. Diseases of Importance in Developing Countries.* Washington, D.C.: National Academy Press.

Jayaraman KS. 1987a. Indian turmoil over planned United States tests of new vaccines. *Nature* 329:94.

―――. 1987b. Vaccines developed in United States to be tested in India. *Nature* 328:287.

―――. 1988. India and the United States agree on vaccine programme. *Nature* 332:198.

Jilg W, Schmidt M, Deinhardt F. 1989. Four year experience with a recombinant hepatitis B vaccine. *Infection* 17:70–76.

Job JS, Halsey NA, Boulos R, et al. 1991. Successful immunization of infants of 6 months of age with high dose Edmonston-Zagreb measles vaccine. *Pediatric Infectious Disease Journal* 10:303–311.

Jones WR, Judd SJ, Ing RMY, et al. 1988. Phase I clinical trial of a World Health Organization birth control vaccine. *Lancet* 1(8598):1295–1298.

Jönsson B, Horisberger B, Bruguera M, Matter L. 1991. Cost-benefit analysis of hepatitis-B vaccination. A computerized decision model for Spain. *International Journal of Technology Assessment in Health Care* 7:379–402.

Jordan P. 1985. *Schistosomiasis. The St. Lucia Project.* Cambridge: Cambridge University Press. 442 pp.

Jordan WS. Jr. 1989. Impediments to the development of additional vaccines: vaccines against important diseases that will not be available in the next decade. *Reviews of Infectious Diseases* 11 Suppl 3: S603–S612.

José MV, Olvera J. 1992. La seroepidemiología de la rubéola en México: Datos y teoría. *Salud Pública de México* 34:328–334.

José MV, Olvera J., Serrano O. 1992. Epidemiología de la rubéola en México. *Salud Pública de México* 34:318–327.

Joyce C. 1986. US exports genetic experiments. *New Scientist* 112:15.

Kambarami RA, Nathoo KJ, Nkrumah FK, Pirie DJ. 1991. Measles epidemic in Harare, Zimbabwe, despite high measles immunization coverage rates. *Bulletin of the World Health Organization* 69:213–219.

Kapikian AZ, Flores J, Hoshino Y, et al. 1986. Rotavirus: the major etiologic agent of severe infantile diarrhea may be controllable by a ''Jennerian'' approach to vaccination. *Journal of Infectious Diseases* 153:815–822.

Kaplan JE, Nelson DB, Schonberger LB, et al. 1984. The effect of immune globulin on the response to trivalent oral poliovirus and yellow fever vaccines. *Bulletin of the World Health Organization* 62:585–590.

Katz SP. 1983. Effects of malnutrition and parasitic infections on the immune response to vaccines. Evaluation of the risks associated with administering vaccines to malnourished children. In NA Halsey, CA de Quadros, eds. *Recent Advances in Immunization.* Pan American Health Organization Scientific Publication 451. 81–89.

Keown CF. 1989. Risk perceptions of Hong Kongese vs. Americans. *Risk Analysis* 9:401–405.

Keusch GT. 1984. Nutrition and Immune Function. In KS Warren, AAF Mahmoud, eds. *Tropical and Geographical Medicine.* New York: McGraw Hill. 212–218.

―――. 1991. Nutritional effects on response of children in developing countries to respiratory tract pathogens: implications for vaccine development. *Reviews of Infectious Diseases* 13 Suppl 6:S486–S491.

Khoman S. 1991. Thailand's patent law and planned amendments. *Bangkok Post.* 30 April.

Kielmann AA. 1977. Weight fluctuations after immunization in a rural preschool child community. *American Journal of Clinical Nutrition* 30:592–598.

Kiepiela P, Coovadia HM, Loening WEK, et al. 1991. Lack of efficacy of the standard potency Edmonston-Zagreb live, attenuated measles vaccine in African infants. *Bulletin of the World Health Organization* 69:221–227.

Kim-Farley RJ, Rutherford G, Litchfield P, et al. 1984. Outbreak of Paralytic Poliomyelitis, Taiwan. *Lancet* 2:1322–1324.

Kim-Farley R, Sokhey J. 1988. A simple screening method for field evaluation of vaccine efficacy. *Journal of Communicable Diseases* 20:32–37.

King M. 1990a. Health is a sustainable state. *Lancet* 336:664–667.

———. 1990b. Public health and the ethics of sustainability. *Tropical and Geographical Medicine* 42:197–206.

Kingsbury DT. 1987. DNA probes in the diagnosis of genetic and infectious diseases. *Trends in Biotechnology* 5:107–111.

Kirchenstein A. 1930. Mit Calmettes Tuberkulose-Impfstoff BCG geimpfte Saüglinge im Riga. *Zeitschrift für Tuberkulose* 57:311.

Kitch EW. 1986. The vaccine dilemma. *Issues in Science and Technology* 2:108–121.

Kliks SC, Nimmanitya S, Nisalak A, Burke DS. 1988. Evidence that maternal dengue antibodies are important in the development of dengue hemorrhagic fever in infants. *American Journal of Tropical Medicine and Hygiene* 38:411–419.

Klipstein FA, Engert RF, Houghten RA. 1986. Immunization of volunteers with a synthetic peptide vaccine for enterotoxigenic *E. coli. Lancet* 1:471–472.

Kristof ND. 1993. China is trying to stifle scandal over reused hypodermic needles. *New York Times,* 31 May.

Kruskall MS, Alper CA, Awdeh Z, et al. 1992. The immune response to hepatitis B vaccine in humans: inheritance patterns in families. *Journal of Experimental Medicine* 175:495–502.

Kyle WS. 1992. Simian retroviruses, poliovaccine, and origin of AIDS. *Lancet* 339:600–601.

Lamm SK. 1988. Use of seroepidemiological tools in the control of tropical diseases. *Tropical Biomedicine* 5(1 Suppl):25–27.

Lancet. 1992. Hepatitis A: a vaccine at last (Editorial) *Lancet* 339:1198–1199.

Landwirth J. 1990. Medical-legal aspects of immunization. Policy and practices. *Pediatric Clinics of North America* 37(3):771–784.

Langille DB, MacDonald J, Sabourin L, Corbett GA. 1988. Contamination of multi-dose vials due to repeat usage of syringes for aspiration. Nova Scotia. *Canada Weekly Diseases Report* Oct. 22 14:193–196.

Last JM. 1983. *A Dictionary of Epidemiology.* New York: Oxford University Press. 114 pp.

Lawrence R, Gershon AA, Holzman R, Steinberg SP, NIAID Varicella Vaccine Collaborative Study Group. 1988. The risk of zoster after varicella vaccination in children with leukemia. *New England Journal of Medicine* 318:543–548.

Lennon JL, Black FL. 1986. Maternally derived measles immunity in era of vaccine-protected mothers. *Journal of Pediatrics* 108:671–676.

Liew FY. 1989. *Vaccination Strategies of Tropical Diseases.* Boca Raton, Fla.: CRC Press.

Lima Guimarães JJ, Fischmann A. 1985. Inequalities in 1980 infant mortality among shantytown residents and nonshantytown residents in the municipality of Porto Alegre,

Rio Grande do Sul, Brazil. *Bulletin of the Pan American Health Organization* 19:235–251.

Luce BR, Elixhauser A. 1990. Estimating costs in the economic evaluation of medical technologies. *International Journal of Technology Assessment in Health Care* 6:57–75.

Lumbiganon P. Panamonta M, Laopaiboon M, et al. 1990. Why are Thai official perinatal and infant mortality rates so low? *International Journal of Epidemiology* 19:997–1000.

Mahoney RT. 1990. Cost of plasma-derived hepatitis B vaccine production. *Vaccine* 8:397–401.

Mariner WK. 1987. Compensation programs for vaccine-related injury abroad: A comparative analysis. *St. Louis University Law Journal* 31:599–654.

Markowitz LE, Sepúlveda J, Díaz-Ortega JL, et al. 1990. Immunization of six month old infants with different doses of Edmonston-Zagreb and Schwartz measles vaccines. *New England Journal of Medicine* 322:580–587.

Marquez PV. 1991. *Selected Issues in Pharmaceuticals. A Background Review for Latin America and the Caribbean.* Belize, C.A.: Pan American Health Organization Regional Office. 204 pp.

Martínez-Palomo A, López-Cervantes M, Freeman P. 1992. The role of vaccine R&D in scientific development of newly industrialized countries. Presented at Conference on Vaccines and Public Health: Assessing Technologies and Global Policies for the Children's Vaccine Initiative, Bethesda, Maryland, November, 1992.

Martuscelli J, Faba J. 1993. Assessment of medical technologies in developing countries: challenges and obstacles. In JR Boldú, JR de la Fuente, eds. *Science Policy in Developing Countries: The Case of Mexico.* Mexico City: Fondo de Cultura Económica. 92–110.

Martyn CN. 1991. Childhood infection and adult disease. *In* GR Bock, J Whelan, eds. *The Childhood Environment and Adult Disease.* CIBA Foundation Symposium 156. 93–108.

Mathias RG, Routley JV. 1985. Paralysis in an immunocompromised adult following oral polio vaccination. *Canadian Medical Association Journal* 132:738–739.

Maynard JE, Kane MA, Hadler SC. 1989. Global control of hepatitis B through vaccination: role of hepatitis B vaccine in the expanded programme on immunization. *Reviews of Infectious Diseases* 11(Suppl 3):S574–S578.

McBean AM, Foster SO, Herrmann KL, Gateff C. 1976. Evaluation of a mass measles immunization campaign in Yaoundé. Cameroun. *Transactions of the Royal Society of Tropical Medicine and Hygiene.* 70:206–212.

McBean AM, Thoms ML, Albrecht P, et al. 1988. Serologic response to oral polio vaccine and enhanced-potency inactivated polio vaccines. *American Journal of Epidemiology* 128:615–628.

McCord C. 1978. Medical technology in developing countries: useful, useless, or harmful? *American Journal of Clinical Nutrition* 31:2301–2313.

McCormick D. 1989. Not as easy as it looked. *Bio/Technology* 7:629.

McGhee JR, Mestecky J, Dertzbaugh MT, et al. 1992. The mucosal immune system: from fundamental concepts to vaccine development. *Vaccine* 10:75–88.

McIntosh K. 1989. Feasible improvements in vaccines in the Expanded Programme on Immunization. *Reviews of Infectious Diseases* 11(Suppl 3):S530–S537.

McMurray DN, Loomis SA, Casazza LJ, Rey H. 1979. Influence of moderate malnutrition on morbidity and antibody response following vaccination with live, attenuated measles virus vaccine. *Bulletin Pan American Health Organization* 13:52–57.

Merson M, Black RE, Sack D, et al. 1980. Maternal cholera immunization and secretory IgA in breast milk. *Lancet* I:931–932.

Milne A, Heydon JL, Hindle RC, Pearce NE. 1989. Prevalence of hepatitis B in children in a high risk New Zealand community, and control using recombinant DNA vaccine. *New Zealand Medical Journal* 102:182–184.

Mitchell LA, Zhang T, Tingle AJ. 1992. Differential responses to rubella virus infection in males and females. *Journal of Infectious Diseases* 166:1258–1265.

Moore PG. 1983. *The Business of Risk.* Cambridge, England: Cambridge University Press. 244 pp.

Morein B. 1990. The iscom, an immunostimulating system. *Immunology letters* 25:281–283.

Mortimer EA, Lepow ML, Gold E, et al. 1981. Long-term follow-up of persons inadvertently inoculated with SV40 as neonates. *New England Journal of Medicine* 305:1517–1518.

Moses LE, Brown BW Jr. 1984. Experiences with evaluating the safety and efficacy of medical technologies. *Annual Review of Public Health* 5:267–292.

Mosley WH, Jamison DT, Henderson DA. 1990. The health sector in developing countries: problems for the 1990s and beyond. *Annual Review of Public Health* 11:335–358.

Moxon ER. 1990. Modern vaccines—the scope of immunisation. *Lancet* 335:448–451.

Murdock GP. 1980. *Theories of Illness. A World Survey.* Pittsburgh Pa.: University of Pittsburgh Press. 127 pp.

Music SI, Schultz MG. 1990. Field epidemiology training programs. New international health resources. JAMA 263:3309–3311.

Nafziger EW. 1984. The *Economics of Developing Countries.* Belmont Calif.: Wadsworth Publishing Co. 496 pp.

Nakajima H. 1989. Priorities and opportunities for international cooperation: experiences in the Western Pacific Region. In MR Reich, E Marui, eds. *International Cooperation for Health.* Dover, Mass.: Auburn Publishing House. 317–331.

Nelson G. 1986. Opening remarks. *Parasitology* 92(Suppl):S3–S5.

Newell KW, Nabarro D. 1989. Reduced infant mortality: a societal indicator, an emotional imperative, or a health objective. *Transactions of the Royal Society of Tropical Medicine and Hygiene* 8:33–35.

NIH. 1992. Research to better understand and prevent measles. *NIH Guide for Grants and Contracts.* National Institutes of Health, Bethesda, Md. 21(21):13–16.

Novotny T, Jennings CE, Doran M, et al. 1988. Measles outbreaks in religious groups exempt from immunization laws. *Public Health Reports* 103:49–54.

Office of Science and Technology Policy. 1988. Proposed model federal policy for protection of human subjects; response to the first biennial report of the President's Commission for the Study of Ethical Problems in Medicine and Behavioral Research. *Federal Register* 51:20204–17.

Ofosu-Amaah S. 1983. The control of measles in tropical Africa: a review of past and present efforts. *Reviews of Infectious Diseases* 5:546–553.

Ogilvie B. 1988. Vaccines: around which corner? *Immunology letters* 19:245–250.

Onoja AL, Adu FD, Tomori O. 1992. Evaluation of measles vaccination programme conducted in two separate health centres. *Vaccine* 10:49–52.

Onorato IM, Markowitz LG, Oxtoby MJ. 1988. Childhood immunization, vaccine-preventable diseases and infection with human immunodeficiency virus. *Pediatric Infectious Diseases Journal* 6:588–595.

Oppenheimer SJ, Gibson FD, Macfarlane SB, et al. 1986. Iron supplementation increases prevalence and effects of malaria: report on clinical studies in Papua New Guinea.

*Transactions of the Royal Society of Tropical Medicine and Hygiene* 80:603–612.

Orenstein WA, Bernier RH, Dondero TJ, et al. 1984. *Field evaluation of vaccine efficacy.* Geneva: World Health Organization. Document EPI/Gen/84/10.rev.2. 43 pp.

———. 1985. Field evaluation of vaccine efficacy. *Bulletin of the World Health Organization* 63:1055–1068.

Orenstein WA, Bernier RH, Hinman AR. 1988. Assessing vaccine efficacy in the field. Further observations. *Epidemiologic Reviews* 10:212–241.

OTA. 1978. *Assessing the efficacy and safety of medical technologies.* Washington D.C.: Office of Technology Assessment. Publication OTA-H-75. 133 pp.

Oyeyinka GO. 1984. Age and sex differences in immunocompetence. *Gerontology* 30:188–195.

Paffenbarger RS Jr. 1988. Contributions of epidemiology to exercise science and cardiovascular health. *Medicine and Science in Sports and Exercise* 20:426–438.

Pappaioanou M, Fishbein DB, Dreesen DW, et al. 1986. Antibody response to preexposure human diploid-cell rabies vaccine given concurrently with chloroquine. *New England Journal of Medicine* 314:280–284.

Pará M. 1965. An outbreak of post-vaccinal rabies (rage de laboratoire) in Fortaleza, Brazil, in 1960. Residual fixed virus as the etiological agent. *Bulletin of the World Health Organization* 33:177–182.

Parker A, Newell KW, Torfs M., Israel E. 1977. Appropiate tools for health care: developing a technology for primary health care and rural develpment. *World Health Organization Chronicle* 31:131–137.

Patriarca PA, Biellik RJ, Sanden G, et al. 1988. Sensitivity and specificity of clinical case definitions for pertussis. *American Journal of Public Health* 78:833–836.

Patriarca PA, Laender F, Palmeira G, et al. 1988. A randomized trial of alternative formulations of oral polio vaccine in Brazil. *Lancet* 1:429–433.

Patriarca PA, Wright PF, John TJ. 1991. Factors affecting the immunogenicity of oral polio vaccine in developing countries: a review. *Reviews of Infectious Diseases* 13:926–39.

Peter G. 1992. Childhood immunizations. *New England Journal of Medicine* 327:1794–1800.

Peltola H, Siitonen A, Kyrönseppä H, et al. 1991. Prevention of travellers' diarrhoea by oral B-subunit/whole cell cholera vaccine. *Lancet* 338:1285–1289.

Petricciani JC, Gracher VP, Sizaret PP, Regan PJ. 1989. Vaccines: obstacles and opportunities from discovery to use. *Reviews of Infectious Diseases* 11(Suppl. 3):S524–S529.

Piachaud D. 1979. The diffusion of medical techniques to less developed countries. *International Journal of Health Services* 9:629–643.

Pillsbury B. 1990. *Immunization: The Behavioral Issues.* [Prepared by International Health and Development Associates] U.S. Agency for International Development, Office of Health. 94 pp.

Pirsig RM. 1974. Zen and the Art of Motorcycle Maintenance. N.Y.: Morrow. 412 pp.

Plotkin SA. 1988. Hell's fire and varicella-vaccine safety. *New England Journal of Medicine* 318:573–575.

Polgar S. 1962. Health and human behavior: areas of interest common to the social and medical sciences. *Current Anthropology* 3:159–205.

Potts M, Paxman JM. 1984. Depo-Provera—ethical issues in its testing and distribution. *Journal of Medical Ethics* 10:9–20.

Programme for Vaccine Development. 1991. *Report of the Eighth Session of the Scientific*

*Advisory Group of Experts (SAGE)*. Geneva: WHO/UNDP Programme for Vaccine Development. Document MIM.PVD/91.8.

Quakyi IA, Otoo LN, Pombo D, et al. 1989. Differential non-responsiveness in humans of candidate *Plasmodium falciparum* vaccine antigens. *American Journal of Tropical Medicine and Hygiene* 41:125–134.

Quinnan GV. 1985. Special safety concerns. In GV Quinnan, ed. *Vaccinia Viruses as Vectors for Vaccine Antigens*. New York: Elsevier. 246–247.

Rahmathullah L, Underwood BA, Thulasiraj RD, et al. 1990. Reduced mortality among children in southern India receiving a small weekly dose of vitamin A. *New England Journal of Medicine* 323:929–935.

Reidy A, Kitching G. 1986. Primary health care: our sacred cow, their white elephant? *Public Administration and Development* 6:425–433.

Reiser SJ. 1988. A perspective on ethical issues in technology assessment. *Health Policy* 9:297–300.

Rimmington A. 1989. Biotechnology falls foul of the environment in the USSR. *Bio/Technology* 7:783–788.

Robbins A. 1990. Modern vaccines. Progress toward vaccines we need and do not have. *Lancet* 335:1436–1438.

Robbins FC. 1977. The demand for human trials in biologicals research. *Bulletin of the World Health Organization* 55 (Supplement 2):73–83.

———. 1988. Polio—historical. In SA Plotkin, EA Mortimer, eds. *Vaccines*. Philadelphia: W.B. Saunders Co. 98–114.

Rosen L. 1977. The emperor's new clothes revisited, or reflections on the pathogenesis of dengue hemorrhagic fever. *American Journal of Tropical Medicine and Hygiene* 25:337–343.

———. 1982. Dengue, an overview. In JS Mackenzie, ed. *Viral Diseases in Southeast Asia and the Western Pacific*. New York.: Academic Press. 484–493.

Rosenthal E. 1993. Doctors weigh the costs of a chicken pox vaccine. *New York Times* 7 July.

Ross DA, Vaughan JP. 1986. Health interview surveys in developing countries: a methodological survey. *Studies in Family Planning* 17:78–94.

Rush B. 1815. An inquiry into the comparative state of medicine between the years 1760–1776 and the year 1809. In *Medical Inquiries and Observations*. 4th ed. Vol. 3:36. Philadelphia: Johnson and Warner.

Russell L. 1986. Is Prevention Better than Cure? Washington D.C.: The Brookings Institution. 134 pp.

Ryan M. 1987. Cheap hepatitis B vaccine divides health experts. *New Scientist* 114:24.

Sabin, AB. 1980. Vaccination against poliomyelitis in economically underdeveloped countries. *Bulletin of the World Health Organization*. 58:141–157.

Sagan LA. 1984. Problems in health measurements for the risk assessor. In PF Ricci, LA Sagan, CG Whipple, eds. *Technological Risk Assessment*. The Hague: Martinus Nijhoff Publishers. 1–29.

Saif LJ. 1985. Passive immunity to cornavirus and rotavirus infections in swine and cattle: enhancement by maternal vaccination. In S Tzipori, ed. *Infectious Diarrhea in the Young*. Amsterdam: Excerpta Medica. 456–467.

Salako LA. 1991. Drug supply in Nigeria. *Journal of Clinical Epidemiology* 44(Suppl. 2):15S–19S.

Saren MA, Brownlie DT. 1983. *A Review of Technology Forecasting Techniques and Their Application*. Bradford (UK): MCB University Press. 74 pp.

Sass HM. 1988. Comparative models and goals for the regulation of human research. In

SF Spicker, I Alon, A de Vries, HT Engelhardt Jr, eds. The Use of Human Beings in Research. Dordrecht: Kleuver Academic Publishers. 47–89.

Sasson A. 1988. *Biotechnologies and Development*. Paris.: United Nations Educational, Scientific, and Cultural Organization. 361 pp.

Schild GC, Assaad F. 1983. New trends in health. Vaccines: the way ahead. *World Health Forum* 4:353–356.

Schneider K. 1989a. Stores bar milk linked to a drug. *New York Times* August 24.

———. 1989b. Vermont resists progress in dairying. *New York Times* August 27.

Schoub BD, Johnson S, McAnerney JM, et al. 1991. Integration of hepatitis B vaccination into rural African primary health care programs. *BMJ* 302:313–316.

Shah K, Nathanson N. 1976. Human exposure to SV40: review and comment. *American Journal of Epidemiology* 103:1–12.

Shapiro CN, Margolis HS. Impact of hepatitis B virus on women and children. *Infectious Disease Clinics of North America* 6:75–96.

Siber GR, Santosham M, Reid GR, et al. 1990. Impaired antibody response to Haemophilus influenzae type b polysaccharide and low IgG2 and IgG4 concentrations in Apache children. *New England Journal of Medicine* 323:1387–1392.

Slovic P, Fischhoff B, Lichtenstein S. 1980. Facts and fears: Understanding perceived risk. In R Schwing, W Albers Jr., eds. *Societal Risk Assessment: How Safe is Safe Enough*. New York: Plenum. 181–216.

Smith PG. 1982. Retrospective assessment of the effectiveness of BCG vaccination against tuberculosis using the case-control method. *Tubercle* 63:23–35.

———. 1988. Epidemiological methods to evaluate vaccine efficacy. *British Medical Bulletin* 44:679–690.

Smith, PG, Morrow RH. 1991. *Methods for Field Trials of Interventions Against Tropical Diseases*. Oxford, UK: Oxford University Press. 326 pp.

Smythe JA, Coppel RL, Day KP, et al. 1991. Structural diversity in the *Plasmodium falciparum* merozoite surface antigen 2. *Proceedings of the National Academy of Sciences U.S.A.* 88:1751–1755.

Soulsby EJL. 1982. The application of the immune response in protozoal infections to immunoprophylaxis in man and animals. *Behring Institut Mitteilungen* 71:104–113.

Steffen R, Woodall JP, Nagel J, Desaules M. 1991. Evaluation of immunization policies for peacekeeping missions. *International Journal of Technology Assessment in Health Care* 7:354–360.

Stetler HC, Mullen JR, Brennan JP, et al. 1987. Monitoring system for adverse effects following immunization. *Vaccine* 5:169–174.

Sutter RW, Patriarca PA, Brogan S, et al. 1991. Outbreak of paralytic poliomyelitis in Oman: evidence for widespread transmission among fully vaccinated children. *Lancet* 338:715–720.

Sutter RW, Patriarca PA, Suleiman AJ, et al. 1992. Attributable risk of DTP (diphtheria and tetanus toxoids and pertussis vaccine) injection in provoking paralytic poliomyelitis during a large outbreak in Oman. *Journal of Infectious Diseases* 165:444–449.

Svennerholm AM, Hanson LA, Holmgren J, et al. 1980. Different secretory immunoglobulin A antibody responses to cholera vaccination in Swedish and Pakistani women. *Infection and Immunity* 30:427–430.

Swartz TA, Skalka P, Gerichter CG, Cockburn WC. 1972. Routine administration of oral polio vaccine in a subtropical area. Factors possibly influencing sero-conversion rates. *Journal of Hygiene* 70:719–726.

Talwar GP, Hingorani V, Kumar S, et al. 1990. Phase I clinical trials with three formulations of anti-human chorionic gonadotropin vaccine. *Contraception* 41:301–316.

Tang D, DeVit M, Johnston SA. 1992. Genetic immunization is a simple method for eliciting an immune response. *Nature* 356:152–154.

Tanouye E. 1992. Merck's "River Blindness" gift hits snags. *Wall Street Journal* Sept. 23.

Task Force on Health Research for Development. 1991. *ENHR Essential National Health Research. A Strategy for Action in Health and Human Development.* Geneva, Switzerland: Palais des Nations. 73 pp.

Taylor D, Laing E. 1989. The economics of vaccines. In Liew FY, ed. *Vaccination Strategies of Tropical Diseases.* Boca Raton, Fla.: CRC Press. 271–282.

Taylor R. 1991. Cholera vaccine draws closer. *Journal of NIH Research* 3:44–45.

Technology Management Group. 1989. Worldwide human vaccine markets to reach $3.7 billion by the year 2000. Seven new vaccines to be commercialized. Company news release. New Haven, Conn. 2 pp.

Touchette N. 1992a. Fact or fiction? HIV and polio vaccines. *Journal of NIH Research* 4(September):40–41.

———. 1992b. Wistar panel disputes polio vaccine-HIV link. *Journal of NIH Research* 4(December):42.

Tsai TF, Bolin RA, Lazvick JS, Miller KD. 1986. Chloroquine does not adversely affect the antibody response to yellow fever vaccine. *Journal of Infectious Diseases* 154:726–727.

UNESCO 1983. *Science, Technology and Development in Asia and the Pacific CASTASIA II.* Paris. UNESCO Science Policy Studies and Documents No. 55. 200 pp.

UNIDO 1992. International code of conduct on biotechnology safety. *Biotechnology and Genetic Engineering Monitor* 37:1.

Vandermissen W. 1992. Availability of quality vaccines: the industrial point of view. *Vaccine* 10:955–957.

van Noordwijk J. 1988. Quality assurance of products manufactured by recombinant DNA technology, and elements of a philosophy. *Arzneimittelforschung* 38:943–947.

Von Magnus P. 1973. Compensation for injuries possibly related to immunization. In FT Perkins, ed. *International Symposium on Vaccine against Communicable Diseases. Symposium Series on Immunobiological Standardization. Vol 22.* London: S. Karger. 325–329.

Wakelin D, Blackwell JM. 1988. *Genetics of Resistance to Bacterial and Parasitic Infection.* London: Taylor and Francis Ltd. 287 pp.

Wall R, Strong L. 1987. Environmental consequences of treating cattle with the antiparasitic drug ivermectin. *Nature* 327:418–421.

Walsh J. 1987. New look at health in developing nations. *Science* 238:746.

Ward J, Brenneman G, Letson GW, et al. 1990. Limited efficacy of Haemophilus influenzae type b conjugate vaccine in Alaska native infants. *New England Journal of Medicine* 323:1393–1401.

Wassilak SGF, Sokhey J. 1991. *Monitoring of Adverse Events Following Immunization in the Expanded Programme on Immunization.* Geneva: WHO. Document WHO/EPPI.GEN/91.2. 29 pp.

Weekly Epidemiological Record. 1992. Expanded Programme on Immunization. Poliomyelitis outbreak. Bulgaria. Geneva: World Health Organization. *Weekly Epidemiological Record* 67:337.

Werth B. 1991. How short is too short? *New York Times Magazine* June 16: pp. 14–17, 28–29, 47.

Whittle HC, Mann G, Eccles M, et al. 1988. Effects of dose and strain of vaccine on success of measles vaccination of infants aged 4–5 months. *Lancet* 1:963–966.

WHO. 1974. Health trends and prospects 1950–2000. *World Health Statistics Report* 27:670–706.

———. 1985. Recombinant vaccinia viruses as live virus vectors for vaccine antigens: memorandum from a WHO/USPHS/NIBSC. *Bulletin of the World Health Organization* 63:471–477.

———. 1986. The use of DNA probes for malaria diagnosis: memorandum from a WHO meeting. *Bulletin of the World Health Organization* 64:641–652.

———. 1988. Basic Vaccinology. A new WHO programme. *World Health Organization Weekly Epidemiologic Record* 63:129–130.

WHO Technical Advisory Group on Viral Hepatitis. 1988. Progress in the control of viral hepatitis: Memorandum from a WHO meeting. *Bulletin of the World Health Organization* 66:443–455.

Wionczek MS. 1979. Science and technology planning in LDCs: Major policy issues. In VL Urquidi, ed. *Science and Technology in Development Planning*. Oxford, U.K.: Pergamon Press. 109–116.

World Bank. 1992. *World Development Report*. New York: Oxford University Press for the World Bank.

Wyatt HV. 1984. The popularity of injections in the Third World: origins and consequences for poliomyelitis. *Social Science and Medicine* 19:911–915.

Youmans GP. 1979. *Tuberculosis*. Philadelphia. Saunders.

Young FE, Nightengale SL, Mitchell W, Beaver LA. 1986. *The United States export drug amendment Act of 1986. An FDA perspective.* 16pp. mimeo. Rockville, Md.: U.S. Food and Drug Administration.

Zilinskas R. 1987. Biotechnology for the developing countries: The role of selected United Nations agencies. *Report to International Cell Research Organization and the United Nations University.* 87 pp.

———. 1989. Biotechnology and the Third World: the missing link between technology and applications. *Genome* (Canada) 31:1046–1054.

Zoppi G, Gasparini R, Mantovanelli F, et al. 1983. Diet and antibody response to vaccinations in healthy infants. *Lancet* 2:11–14.

Zuckerman AJ. 1985. Controversies in immunization against hepatitis B. *Hepatology* 5:1227–1230.

# Index

*f* = figure; *t* = table